Sum Thing is Up: A Practical Guide to Dyscalculia

Tips, Tools, and Resources for Parents, Educators, and Anyone Affected by Dyscalculia

Dr. Sandra Elliott

Sum Thing is Up: A Practical Guide to Dyscalculia by Dr. Sandra Elliott

First edition 2025.

ISBN: (paperback) 978-1-966551-23-2
ISBN: (eBook-Kindle) 978-1-966551-24-9

TouchMath Acquisition LLC
2 N Nevada Ave, Ste 1200
Colorado Springs, CO
80903-1702

www.touchmath.com

AUTHOR'S NOTE

Writing about science and teaching for my education colleagues and parents requires balancing brevity with detail and depth with readability. I've done my best to find the perfect balance, being able to explain just enough of the science to establish credibility for an idea, avoid academic jargon, and make it accessible to everyone. I'll let you be the judge of whether or not I was successful and that you find this book useful in supporting students with dyscalculia or just struggling with math. And if you read only one chapter, read Chapter 8, the case studies of adults who did not know their math problems and feelings of "stupidity" were due to the disability, dyscalculia. There are so many lessons to be learned by listening to their voices.

PS- I have yet to learn how to use any of the AI tools, so you are reading my words. I'm not going to promise that for the next book.

FOREWARD

By Dwight Jones, CEO of Education Partners

As the Founder of Education Partners, a boutique advisement group to states and school district and a former state commissioner of education and superintendent of schools, I get to see just how important math and literacy are to being a successful adult and to our nation's health as a whole. I also see how difficult it can be for those with disabilities to be successful if they are not identified and receive the support needed to thrive as adults.

Our nation is very familiar with dyslexia and nearly everyone can describe that it is a learning disability that interferes with the reading process and that it does not need to prevent a student of successfully proceeding down any career path they choose. The same cannot be said for dyscalculia and while we may debate which skill, literacy or numeracy is more important, mathematics is an essential life skill and dyscalculia shouldn't prevent a child from being successful at the career or job of their dreams.

Dr. Sandra Elliott and I have a long history of debating which is more important. She has helped me recognize that dyscalculia is a grossly underrecognized and diagnosed disability and one that needs our attention as parents, educators, and as a nation. I have been able to listen to her emotional appeal to acknowledge dyscalculia, how to recognize it, and how we already have the tools to provide interventions at the start of PreK or even kindergarten that are successful. She has made the case that it is time to tackle this disability just like we did with dyslexia 20 years ago and have adults define it as easily as they do dyslexia. As a former Exceptional Education teacher, 5-time principal, district administrator, and as Chief Academic Officer she knows what she is talking about. She led the development of the free screener for dyscalculia, DySc, and has written and presented around the world on the subject.

Reading this book is like sitting in the crowd as she presents the latest research about what causes dyscalculia, how to spot, screen, and diagnose it as early as age 3, the research and evidence based teaching and learner strategies that enable these students to master mathematics as well as their non-disabled peers. She has taken a complex subject and made it understandable and more importantly, actionable for the reader. It should be referred again and again to teachers and parents whose students or children are struggling with math but show no evidence of a cognitive reason for that struggle. It provides a variety of tools and actions that can be implemented immediately to support a student who has dyscalculia or is just struggling with math. In fact, if enough teachers and parents read this and implement her recommendations, we would see more students viewing math as a ticket to the careers of their dreams and especially those that are STEM and math related careers as well recognition that numeracy is as important as literacy for the successful adult.

Dwight Jones was the former Commissioner of Education, State of Colorado and the Superintendent for Clark County School District, Nevada one of the top 10 largest school districts in the country.

PREFACE

Our proficiency at using basic and advanced math is critical in determining our futures. Mathematical skills are important when looking for and holding a job, having a career, and functioning in everyday life; think computers, smartphones, our bank and credit cards, and keeping track of health and prescription information. Most jobs require us to use math. The fields of science, technology, engineering, medicine, and others demand an even more advanced understanding of mathematical concepts. But current facts show an alarming number of individuals with the inability to perform everyday math tasks.

In May of 2022, the U.S. Institute of Education Sciences estimated that 33% or 70 million+ of Americans had difficulty doing calculations with whole numbers, comprehending living on a budget, understanding how to read figures in a simple graph, estimating the tip for a restaurant bill, or measuring ingredients when preparing a meal. This includes the math skills needed to perform jobs in today's world. The 2022 National Assessment of Educational Progress highlights a nearly universal decrease in average math scores for both fourth (5-point decrease) and eighth grades (8-point decrease) compared to scores in 2019 (Duran et al., 2020; National Center for Education Statistics, 2022).

In the last 20 years, there has been a push to address illiteracy through early identification with universal screeners, but the same focus has not been there for innumeracy. Even with the increased need for math skills due to explosion of STEM and technology-focused jobs, and the recognition that most jobs and daily life activities require math there have not been national campaigns to address the problem of so many children graduating high school and NOT being proficient in math (Bryant, 2008; De Visscher, 2018; National Center for Education Statistics, 2022). That same proficiency in math is also necessary for successfully completing career and college classes.

Early math skills strongly predict later math achievement (Kroesbergen et al., 2022; Mattison et al., 2023). During an individual's school life, poor math skills will negatively impact a child's academic achievement, usually preventing or at least decreasing the opportunity to access the advanced math classes that are required for a multitude of career paths. This lack of access increases the likelihood of dropping out of school and is a contributing factor to unemployment (or under-employment), in adulthood. The impact of innumeracy on everyday adult life is clear in bankruptcies, lower-wage job opportunities, unemployment, and a higher probability of physical and mental health issues, including anxiety and suicide (Dowker et al., 2016; Duncan et al., 2007; Geary, 2011; Jacobsen, 2020). Without effective interventions those who struggle with math, a necessary life skill, begin to feel a lack of ability, confidence, and an increase in anxiety. These in turn worsen the difficulties in mathematics learning, with the individual avoiding the subject of math and in many cases the choice of careers that appear to require advanced math. You will see as you read Part III of this book, where I share the interviews, I have completed to date of adults who were diagnosed after high school with dyscalculia, that all of them felt that due to their math skills, their dream career and in several instances a college education was out of the question.

There is also a cumulative economic impact estimated in the billions of dollars in lost revenue and increased costs in social support when large percentages of a population are unemployed or underemployed, due to their inability to perform the normal math required of an adult. According to OECD (Kankaras, 2016) and IES (2022) estimates, the share of adults at or below Level 1–the ability to do mathematical processes such as counting, sorting, and basic arithmetic operations–ranged from

61.9% in Chile to 8.1% in Japan, with the United States at around 33%. That translates to one-third of the U.S. workforce not able to do math that a proficient 3rd or 4th grader can perform. Why are these figures, national and international, important? Because workforce participation enables economic growth, stability and prosperity for everyone.

The disparities between students with and without Individualized Education Programs (IEP)s were even more disturbing. On the 2022 National Assessment of Education Progress (NAEP), 84% of 4th grade students with IEPs scored below proficient compared to 61% without IEPs (Hroncich, 2022). In eighth grade, 93 % of students with IEPs scored below proficient compared to 70% without an IEPs.

How students with different disabilities perform

Many factors can cause poor numeracy skills; frequent moves, excessive absenteeism, poor instruction, intellectual and emotional disabilities, and motivation, to name a few. However, one commonly overlooked cause is the neurodevelopmental learning disorder estimated to affect 5-7% of the worldwide population. This is the currently greatly underdiagnosed learning disability specific to math called dyscalculia. If diagnosed, it can be addressed, and individuals with this disability can demonstrate numeracy skills equivalent to their non-dyscalculic peers (Haberstroh & Schulte-Korne, 2019; Kuhl et al., 2021). Early screening for dyscalculia and other math difficulties, seeking a formal diagnosis, and then assigning appropriate supports and evidence-based interventions is critical to successfully addressing the short- and long-term impacts of this learning disorder (Aquil & Ariffin, 2020).

Although there has been an increase in knowledge and interest in the disorder, it has been estimated that the ratio of research on dyslexia compared to dyscalculia could be 14:1 (Devine, 2018; Haberstroh & Schulte-Korne, 2019; Price & Ansari., 2013). The disparity in research between the disorders leads to a lack of general knowledge by the public about dyscalculia and a dramatic difference in the numbers of students identified and diagnosed as dyscalculic. There are fewer screening and diagnostic tools (Jaya, 2009), which also contributes to the underdiagnosis problem, but the prevalence of dyscalculia is similar to dyslexia (Menon et al., 2020; Price & Ansari, 2013; Santos et al., 2022). Add to that an estimated comorbidity of 35-70% with other neurological disabilities, including dyslexia, and suddenly there are a staggering number of students who are at risk of failing math when early identification and intervention can mitigate the problem in many cases (Kisler et al., 2021; Litkowski et al., 2020).

Even using a conservative estimate level of 5% of the 55+ million students in US public and private schools in the fall of 2022, that would amount to approximately 2.75M students (about the population of Kansas) with some degree of math learning disability. The most recent IES data is for the 2020-21 school year, which records the number of students served in K-12 schools with disabilities as 6.7 million, with 1.2 million diagnosed as Specific Learning Disability (SLD), the majority diagnosed with SLD-reading compared to the other SLD types-dysgraphia, dyspraxia, and dyscalculia (Kurth & Jackson, 2022). It must be noted that not all students with a learning disability are diagnosed or need an IEP if they are diagnosed. If their academic performance is not impacted by their disability, it may never be found explaining some of the disparity in numbers noted above.

Why does dyscalculia usually go undiagnosed? There are several possible reasons. Early national and international literacy campaigns more than 20 years ago focused on dyslexia which led to more awareness and, as a result, more tools and actions for students struggling with reading and being diagnosed with dyslexia. It is also more socially and culturally acceptable to struggle with math compared to reading. Have you ever heard an individual say, "I'm bad at reading' or "I don't like to read?" Probably not as it is not socially acceptable. The same does not go for the phrase, "I am bad at math." That phrase usually gets sympathetic head nods and comments of "It was not my favorite subject in school." It does not cause the same anxiety and demand for diagnosis of the problem and support for the child as does the phrase, "I am bad at reading". Research has also been grossly underfunded for dyscalculia compared to dyslexia resulting in a comparative lack of validated and reliable assessments to identify students with dyscalculia, resulting in less identification (Geary, 2011; Halberda et al., 2008; Price & Ansari, 2013).

Like reading, there are foundational skills upon which most mathematics is based- sorting and classifying, recognizing abstract symbols (numerals) and their concrete counterparts, counting to ten, the operations of addition, subtraction, multiplication and division. Children begin learning these skills and concepts before the age of three. Conceptual understanding and procedural fluency of the basic operations in first and second grade underlie success in Operations & Algebraic Thinking (OA) and Numbers in Base Ten (NBT), which are the basis for more advanced math's (Fuchs et al., 2012). In the various U.S. State Mathematics Standards, students begin learning number sense and base ten in pre-K and kindergarten (ERIC, 2022; New York State Education Department, 2017). They are expected to master addition and subtraction by the end of 2nd grade, multiplication by the end of 3rd grade and multi-digit operations by the end of 4th grade. Failure to master these can lead to failure to master the more advanced math concepts and skills needed for middle school, high school, and job success. If you look at the performance of students with disabilities on the NAEP 2022 (Hroncich, 2022) you must also ask what percentage of these students go on to trade and technical schools or college? Do they choose to drop out before completing high school or further training? The dismal 2023 employment rate of 22.5% for disabled adults compared to that of 65% for adults without disabilities seems to indicate that it is not many (De Maria & McLaren, 2024.)

In math, most screening and diagnosis processes are aimed at the student already in school, usually 3rd grade and beyond, and typically showing signs of struggling. At that point, multiple years of potential support have been lost. It is not uncommon for the request for help to happen when students are two or more years behind and struggling to master basic math foundational skills and have not been diagnosed as having a disability. This is years too late and does not need to be the norm any longer.

Just as not everyone who struggles with reading has dyslexia, the low number of students scoring proficient or above on math suggests there are likely to be undiagnosed math learning disabilities and many students struggling with math but not having a disability. The large numbers of students whose test scores are below proficient underscore the importance of identifying and addressing the math difficulties of all students, especially those with disabilities.

Students with dyscalculia or the specific learning disability-math struggle and can have significant difficulties with composing and decomposing numbers, retrieval of number combinations, conceptualizing and applying number concepts, limited working memory, and inefficient problem-solving strategies. However, the research coming out of educational neuroscience and the use of the fMRI and other advanced tools support being able to identify the risk for dyscalculia in early childhood, well before 3rd grade, and successfully intervene, enabling students with dyscalculia to achieve age-appropriate math achievement levels (Bailey, et al.; 2020; Brendefur, 2018; Bugden et al., 2020; Morsanyi et al., 2018). Early identification in the years preceding 3rd grade and intervention in the specific math problem areas is effective and yields the best prognosis for students (Haberstroh & Schulte-Korne, 2019).

We know that early math skills strongly predict later math achievement (Kroesbergen et al., 2022; Mattison et al., 2023). The impact of poor math mastery supports the need for developing and widely using screeners to identify students who are struggling and determine interventions that address the math deficits and struggles. While much progress has been made with screeners for dyslexia and addressing the foundational skills for reading, this is not the case for numeracy issues (Al Otaiba, 2020). The adoption of easy-to-use universal screeners, dyscalculia screeners, and other data collection tools such as surveys and curriculum-based measures that collect information from educators and guardians, is a necessary next step. These tools can quickly assess performance–where a child is functioning in comparison to grade level expectations or the primary indicators of possible dyscalculia–and they can also provide action and intervention plans for adults to use while the formal diagnosis is undertaken.

This book is an attempt to combine what we know from the sciences- educational, cognitive and neuro- about dyscalculia and the most effective ways to support the development of math mastery and the strategies that those with dyscalculia say enable them to successfully take part personally and economically in society. It is written for those that educate students, have children with dyscalculia, or have an interest in dyscalculia so they can access and use the knowledge that is so much more readily available today.

The book has 5 parts with multiple chapters in each part. Part I shares the currently accepted definition of dyscalculia and its descriptors as well as the highlights of current neurological science around its causes and what teachers would actually see when the student is struggling in class. Part II describes the current state of screening and diagnosis and the components of a good screener. Part III provides information and recommendations from a series of interviews that were held with individuals diagnosed with dyscalculia and what they do to navigate the world while having a math disability. Part IV delves into what supports and accommodations need to be provided for children with an emphasis on the social emotional aspects of knowing that something is not right. When one believes they are going to fail, they typically do and not dealing head on with math anxiety only worsens the problem. Part V focuses on what research and evidence show are the key interventions for students with dyscalculia and activities that can be easily done today by teachers and parents. The book concludes with suggestions for other resources and websites you can access for more information. And of course, for those of you like me, the pages of references I have sourced to support what I have written.

INTRODUCTION

Why another book on math? Because we have yet to treat poor math knowledge-innumeracy- with the same respect and fear we give illiteracy. I don't have to define illiteracy it is a widely understood term unlike its counterpart innumeracy. The same goes for the term dyslexia. I challenge you to ask a friend or a stranger if they know what dyslexia is? Out of 100 asks, you'll probably get 99 yeses and a decent statement of its relationship to reading. Now ask if they know the name of the math disability. You can also wonder aloud if there is one for math. Typically, their head will tilt, and they will say they never really thought about it and ask you "what is the name of the math disability?"

This book is designed to decrease some of that ignorance as it addresses the widely undiagnosed and often overlooked specific learning disability called dyscalculia. It provides an accessible and easy-to-understand resource for parents and educators supporting students with dyscalculia, as well as those who may have it themselves. It may not always use the most scientific language or precise math terminology as this is a layperson's book in an attempt to get the main ideas and interventions in front of as many people as possible. By exploring current research from the neurological, cognitive, and educational sciences, we can better understand the causes of dyscalculia, emphasize the importance of early screening, and explore effective interventions for this disability.

In addition, this book incorporates interviews I conducted in 2024 with adults who shared their personal experiences coping with dyscalculia.

They offer their insights into what they would tell their 14-year-old selves, as well as advice for you, their parents and teachers. The responses from these interviews will hopefully inform the actions you can take to support our students, and they may even inspire future research on effective interventions.

Throughout the book, I have attempted to provide examples of how parents, educators, and individuals impacted by dyscalculia can address math-related difficulties and work towards math mastery. I have also shared easily accessible resources.

Mathematics is a critical body of knowledge that impacts our self-worth and our economic success as individuals and as a country. By deepening our understanding of dyscalculia and addressing math struggles, I aim to contribute to a body of knowledge and practice that reduces the number of individuals who feel frustrated or disconnected from math concepts. My hope and that of my colleagues at TouchMath is to empower caring teachers and adults with the understanding and skills needed to effectively help individuals learn math and emphasize the importance of treating math struggles with the same respect and attention as we do with illiteracy.

If you have been reading numerous books on dyscalculia, this book aims to reinforce your knowledge and practices with research-driven, evidence-based statements. I hope that this information will assist in early identification, instruction, and accommodations for the estimated 5-7% of the general population of students that have dyscalculia. I hope to provide ideas that are easy to implement on how to assist the other 15% with some of the risk factors. Additionally, this book can serve as a valuable resource for professional development for teachers and provide guidance and activities for parents seeking practical supports.

Together, let's work towards creating a more inclusive and supportive environment where students with dyscalculia can thrive and succeed in the world of mathematics and not believe as my interviewees did, that they can't grow up to have the career of their dreams. As so many have said, but it is very appropriate here, that is wrong in so many ways.

PART I

In this chapter you will learn:

- How to pronounce dyscalculia
- What dyscalculia is and is not
- The long history of dyscalculia
- Causes of dyscalculia
- What dyscalculia looks like in children and adults

"I did not get it; it felt like there was something wrong with me in math and it just made me feel broken."

Subject Carrie interview

Chapter One

DYSCALCULIA: WHAT IT IS AND WHAT IT IS NOT

Pronouncing Dyscalculia

According to the Cambridge and Marriam Webster dictionary the pronunciation is the same except for one version has a clearly American accent and the other a British accent. Go to touchmath.com/dyscalculia-101/ to hear the two most used pronunciations.

Phonetically it is spelled /ˌdɪs.kælˈkjuː.li.ə/ and is pronounced dis-cal-cue-lee-uh.

What is Dyscalculia

Our ability to learn mathematics is not an either-or situation, we all learn math and use it. With that being said, how well we learn and then utilize math concepts and procedures to solve problems in our daily lives is a result of the interactions of our genetics (our cognitive abilities and the physical structures of our brains), and environmental factors such as school, home, exposure to math careers and activities. Nearly everyone is familiar with the bell curve and the fact that we each fall somewhere on it depending upon the topic we are discussing, math, playing baseball, having perfect pitch. We are a beautiful 3-dimensional topographical map of abilities and potential. Where we each fall in our success at utilizing math is somewhere along that continuum called the normal distribution or bell curve. Some of us will be the far right of the curve with extremely developed math knowledge, those with a Ph.D. in advanced mathematics; to extreme struggle; those with dyscalculia where the numbers make no sense and something like 2+3 needs to be counted on fingers.

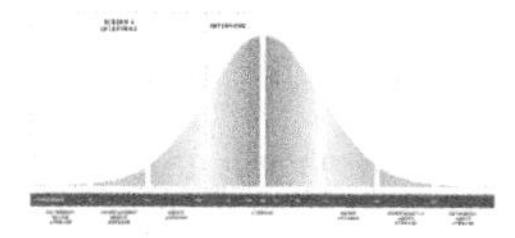

Fortunately, our innate ability to master and use math is not set in stone. We are not just a static point of the bell curve but rather a moving dot that is able to move along it based on how hard we are willing to work, how much support we get from our teachers and families, and the amount of time we have invested in our learning math on any given day. For those with dyscalculia this means the difference in not being able to figure out if they a mortgage is affordable on their current salary to being a paramedic who uses math daily to make life and death decisions about dosages that need to be administered for those in cardiac arrest.

So, what is dyscalculia? To understand that it's helpful to understand how specific learning disabilities came into being. The history of dyscalculia isn't well documented before 1969, when it became written into U.S. federal law. Part of this may be due to the nature of education globally. Before the 1900's literacy and numeracy were limited to a very small group of the population as most individuals did not receive what we think of as a traditional education.

As neuroscience began to mature in the early 1900's, and more of the population learned to read and do math, studies began to separate the language and writing issues from the math deficits. In 1924, Josef Gerstmann, an Austrian neuroscientist, described a condition that included finger agnosia (the inability to name, touch or move a specific finger when asked), dysgraphia (a learning disability that affects the ability to write), and dyscalculia (Altabakhi, et al., 2023). Margaret Reinhold published a paper in 1951 that more clearly described the struggles that adults had with understanding the connection of the operational symbols (+ - x /) and the procedures of addition, subtraction, multiplications and division. She also described how these individuals had difficulty drawing and labeling clocks or grids (as cited in Wall, 2022).

The major change in the U.S. came about a little over 50 years ago in 1969, when the term "specific learning disability" (SLD) was first written into the Specific Learning Disabilities Act (PL 91-230) as a disability that can interfere with academic performance and life. SLD was defined as a child not meeting state-approved age or grade level standards in one or more of the major domains, reading, mathematics, and written expression. The American Psychiatric Association changed the diagnostic criteria for SLD so that it included all three learning disorders. SLDs are "typically diagnosed in early school-aged children, although they may not be recognized until adulthood. They are characterized by a persistent impairment in at least one of three major areas: reading, written expression, and/or math." (APA, 2020). Currently SLD students represent the largest percentage of students being served in special education classes or having IEPs. (Grigorenko, 2020).

Current estimates are that as much as 5-7% of the world's population may have dyscalculia to the extent that it significantly impacts their lives while another 13-15% have some of the factors. This means approximately 1 in 15 children, adolescents, and adults (Butterworth, 2003; Haberstroh & Schulte-Korne, 2019; Menon et al., 2021; Price & Ansari, 2013; Ustun, 2021), or that there is a high probability that one or more students in every classroom may have dyscalculia.

In the 50 years that have passed since specific learning disabilities were written into law, research has provided insight into the causes, manifestations, and interventions for SLD but not in equivalent amounts for literacy, numeracy, writing and other areas. Reading has received a disproportionate share of the research funds with current estimates being that for every dollar spent researching dyscalculia, 14 are spent on dyslexia (Devine, 2018; Haberstroh & Schulte-Korne, 2019; Price & Ansari., 2013) Based on undocumented conversations I've had the figures may be improving with estimates possibly being in the range of 10:1 for research funds dedicated to dyslexia vs dyscalculia.

Reasons for this disparity in research funding may be due to our nation seeing literacy as more important than numeracy and the common myth that math is something you are either born "good at or not". A recent November 2024 poll conducted by the author during a dyscalculia workshop for one of the top 20 US school districts had two-thirds of the nearly 190 participants indicating that reading was more important than math in day-to-day success and careers.

The relatively recent integration of the fields of neuroscience, psychology, genetics, and educational science has led to new knowledge about these learning disabilities. With new technologies we are now able to see how our brains are wired and actually process math. This new knowledge has led to increased research into the other learning disabilities which should improve our ability to treat them as it did with dyslexia.

Definitions of Dyscalculia

Now for the definition and there are a number, but you can see the similarities for yourself. Dyscalculia or the specific learning disorder for mathematics is a neuro-developmental disorder with persistent difficulties in acquiring number-related skills, which cannot be attributed to intellectual disabilities or neurological disorders according to the Diagnostic and Statistical Manual for Mental Disorders, 5th Edition or the DSM-5 as it will be called going forward (American Psychiatric Association, 2022). According to the DSM-5, "...dyscalculia is a term used to describe difficulties learning number-related concepts or using the symbols and functions to perform math calculations. Problems with math can include difficulties with number sense, memorizing math facts, math calculations, math reasoning, and math problem solving" (American Psychiatric Association, 2022). Our educational and legal system in the United States uses the term disability, which entitles the individual to recognized status as a person with a disability and therefore entitled to special services and accommodations in the educational system (Üstün et al., 2021).

In simple terms, dyscalculia impacts a child's ability to acquire arithmetical skills due to a neurodevelopmental disorder, not intelligence or schooling. It usually manifests in the early school years, frequently as children learn number sense and base ten concepts and shows persistence over at least six months (Clements et al, 2013). This difficulty with non-symbolic and symbolic numerical processing means the child will struggle to quickly estimate which is more or less, remember the steps in a multistep math problem or perform the mental math required to solve simple calculations without writing out the process. There is also a reliance on verbal or physical strategies during counting, such as finger counting or counting aloud (Grant et al.; 2020; Lopes-Silva et al., 2016). Children typically lag their peers, which is unexpected based on their intelligence and performance in other school subjects; they struggle with one or more of the four mathematical domains identified in the DSM-5: number sense, memorization of arithmetic facts, accurate and fluent calculation, and accurate mathematical reasoning. Functionally, there are also significant differences in the areas of the brain that are activated in order to perform mathematical tasks. Structural differences in white and grey matter have also been noted (Fengjuan & Jamaludin, 2023). These students can also have a diverse range of other difficulties with mathematics; from math anxiety to math avoidance, following procedures, or being slow to solve problems. It is unexpected, occurs across all ages, and does not appear to be gender-specific (APA, 2022; Ashraf et al., 2021; Butterworth, 2003; Butterworth, 2012; Geary, 2013; Price & Ansari, 2013; Ustun, 2021).

The International Classification of Diseases (ICD-11) has a very similar description of dyscalculia -- significant and persistent difficulties in learning academic skills, which will include reading, writing, or arithmetic. The individual's performance in the affected academic skill(s) is markedly below what would be expected for chronological age and general level of intellectual functioning, resulting in significant impairment in the individual's academic or occupational functioning

(Burns et al, 2010; Carey, 2009). The developmental learning disorder first manifests when academic skills are taught during the early school years. For the individual diagnosed with dyscalculia (6A03.02), the 'Learning difficulties are manifested in impairments in mathematical skills such as number sense, memorization of number facts, accurate calculation, fluent calculation, accurate mathematical reasoning.

Other definitions such as the British Dyslexia Association's; "Dyscalculia is a specific and persistent difficulty in understanding numbers which can lead to a diverse range of difficulties with mathematics. It will be unexpected in relation to age, level of education and experience and occurs across all ages and abilities." (2024 BDA) and the SpLD Assessment Standards Committee in the UK are in general alignment with the APA (SASC, 2019). There is growing consensus that it exists on a continuum and learner profiles will vary based on the areas of the brain that are impacted from severe dysfunctionality to those that do not dramatically impact academic achievement and day to day functionality.

A favorite definition, although much older than most, sums it up pretty well. The UK's National Numeracy Strategy described dyscalculia in simple terms that still hold true today although we know so much more about its causes and what we need to do. "Dyscalculia is a condition that affects the ability to acquire arithmetical skills. Dyscalculic learners may have difficulty understanding simple number concepts, lack an intuitive grasp of numbers, and have problems learning number facts and procedures. Even if they produce a correct answer or use a correct method, they may do so mechanically and without confidence." (DfES, 2001).

Other terms have also been used interchangeably when describing the struggle to learn and remember arithmetic facts and perform basic operations. These have included mathematical learning difficulties, math disorders, math disabilities, mathematical dyslexia, math learning disability, dyscalculia, and developmental dyscalculia (Mahmud et al., 2020; Morsanyi, 2018; Van Luit & Toll, 2018).

Although the formal definitions do not include an assessment score on standardized tests, students are usually identified for screening or for the diagnosis process based on an arbitrary cut-off score to help determine who is eligible for screening or diagnosis. The score is commonly generated via a standardized state assessment and can range from the 2nd to the 40th percentile depending upon the school or district (Kroesbergen et al., 2022). The most frequently mentioned percentile is the 25th, where a strong possibility of a learning disability will exist as there is a significant struggle at this level and the student needs further support (Cheng et al, 2018). Another form of data that can be used as an indicator of potential dyscalculia is when a student is 2 or more grade levels behind in math. The widespread use of universal screeners for reading and math make this an easy tool for determining if further investigation into the causes of poor math achievement is warranted.

Chapter Two

THE COMPONENTS OF DYSCALCULIA

Dyscalculia is not temporary, and there is no cure, but it is also not always an issue in academic development. Because it is not due to a disorder of intellectual development, sensory impairment (vision or hearing), neurological or motor disorder, lack of availability of education, lack of proficiency in the language of academic instruction, or psychosocial adversity; individuals with dyscalculia need not be limited in their academic or career success. There are no treatments or medications that can 'cure' dyscalculia. What is important to know is that research supports the idea that interventions can improve the understanding of mathematical concepts and procedural fluency; and there are accommodations and modifications in school and life that can mitigate the effects of the disorder (APA, 2022; Bailey et al.; 2020; Bugden et al., 2020; Kuhl et al., 2021).

What areas of math will be a struggle for individuals with Dyscalculia?

Dyscalculia is not due to lack of intelligence; IQ and performance in other subject areas is typically average or above and some math skills are present, but they are typically poor. This is especially true for number sense, math facts and calculations. As noted earlier, when diagnosing dyscalculia, the DSM-5 requires that number sense, memorization of arithmetic facts, accurate and fluent calculation, and accurate mathematical reasoning be the areas that are evaluated as they are the most common areas of struggle. The cognitive sciences and the neurosciences describe these deficits as domain-specific and they have been found to be predictive of later struggles in math achievement and dyscalculia. But not every child will display weaknesses in all the areas as there are as many differences among that 5-7% with dyscalculia as there are between children with dyscalculia and those without dyscalculia (Kroesbergen et al., 2022; Menon et al., 2020). Other areas of difficulty are more general and can be found in individuals with ADHD, dyslexia, ASD, etc. These are called domain-general and include components of executive functioning, attention, phonological processing and inhibition control.

Let's go into more detail so you can see some of the nuances. The individual who has dyscalculia typically has problems with:

Number sense- The student struggles with recognizing and understanding quantities, spoken and written number words and Arabic numerals. There are also difficulties mapping between them, knowing that seven and 7 represent the same number (Geary, 2013). Number sense is the start to how we master math and make sense of it, and it has a number of components that are necessary for children to master. These include everything from understanding what a number is, counting, number patterns, subitizing, using a number line, arithmetic fact retrieval, and problem-solving operations. Examples include:

Understanding that 3 is different from 6 as the child has not attached a quantity to the number. The numerals 3 or 6 are simply squiggles that have no meaning as they do not see a picture in their heads of 3 items. And the words three and six are simply words and not attached to the symbols 3 or 6 nor the quantities. These difficulties can be used to screen for dyscalculia before grade 3 (Menon, 2020).

Counting is not onetwothreesix, counting is a series of distinct words each representing a unique quantity; one, two, three, four, five.

Subitizing is included in this as the student has trouble seeing a plate with 3 cookies and being able to tell you that there are 3 cookies without counting them.

Estimation is also frequently an issue. When asked to estimate if the sum of 45 + 54 is closer to 100 than to 75, they cannot understand that both are closer to 50 so it must be 100, rather they may take the long route and actually perform the arithmetic calculation of 45 + 54 to get 99.

Counting backwards is also an issue as it would be similar to us trying to recite our phone numbers backwards, most of us find it very difficult to do from memory.

Pattern recognition is hard to develop. 2+2 = 4, 3+3 = 6, 4+4 = 8 are a doubling pattern but not to someone with dyscalculia.

Place value becomes another challenge as the position of the number in a string determines its value. The number 7 in 737 has 2 completely different values totally based on it being in the one's column but also in the hundred's column. This also comes into play when asked to transcribe numbers such as in a word problem or dollar amount. One hundred thirteen becomes 10013 and not 113.

Time is also a difficulty not just for using an analog clock where the numerals may be not shown or only partially shown, but also for the estimation of how time passes. The difference between 10 minutes and just a minute, much less an hour can be nonexistent.

Memorization of Arithmetic Facts – Fluency and accuracy, being able to instantly recall that 7x6=42 is a skill that is expected of all students. And it serves a purpose, freeing up working memory. For those with dyscalculia it can be problematic. Examples include:

The student memorized the 4's table yesterday but cannot recall it today or is slower than expected recalling it. This begins as early as kindergarten and 1st grade and can be used as an early indicator of potential dyscalculia.

The student uses immature strategies to process math facts such as finger counting or counting all. This slower speed in completing problems leads to a significant time difference between a typical student and the student with dyscalculia (Mahmud et al., 2020; Price & Ansari,

2013) and if not accounted for on an IEP can dramatically impact performance on high stakes tests.

The student may also struggle with subitizing. When they see a pair of dice or a playing card with the distinctive arrangement of 5 dots, they do not recognize it as 5 immediately but may need to actually count the dots in order to know that it is 5.

There are also struggles with decomposing and composing numbers, 7 is 1+6 but not 5+2, understanding how to make tens, the decimal system, etc.

Accurate and Fluent calculation – Being able to use the basic operations of addition, subtraction, multiplication and division is being able to use the math facts and work out a math problem. Examples of struggles with calculation include:

Students with dyscalculia may need more time to solve a problem and make more mistakes, they may use their fingers, hash marks, objects, calculators, math tables etc. to solve a single or multi-step problem (Mahmud et al., 2020; Szardenings et al., 2018).

Guessing is also common and usually not related to any recognizable rule.

The student will know the fact yesterday but does not know it today or could fail to recall it when the context changed, they can solve the word problem but not when completing a set of practice problems.

Accurate mathematical reasoning – A student may be able to memorize the facts and calculate but must also be able to reason and make sense of a problem. To solve a problem means knowing the rules or applying them and whether the answer makes sense. Students with dyscalculia show difficulties with:

A student has difficulty with quantity and magnitude judgment (estimating numerosity) and manipulation using abstract representations or numerals to reason with when presented with problems. The issue is one of mapping the numeral or symbol to the actual representation of that quantity or attaching a numeral to the number of items in a set. A student will be asked which is greater- two numerals (symbolic) or 2 sets of items (non-symbolic) or a number and a set of items which are displayed simultaneously, and the student will have difficulty distinguishing between the two.

Another example is number sequencing which includes skip counting intervals with increasing or decreasing numbers (Lewis et al., 2022; Menon et al., 2020).

The difficulty with magnitude judgment also includes estimating the duration of time and spatial dimensions so being late or struggling with the measurements in a recipe are not unusual. Scheduling meetings, being on time, using calendars, driving at the speed limit and other day-to-day tasks can also be impacted.

This combination of struggles with number sense, mathematical facts, calculation, and reasoning impacts the ability to successfully master later and more complex math skills. They limit the ability to generalize the more basic math skills to everyday problems and new contexts.

In addition to the four-math domain-specific weaknesses there are domain-general difficulties that stem from other regions of the brain involved in doing mathematics. These include skills such as visuospatial and phonological skills, processing speed and executive functioning. Executive functions skills include working memory and attention, time management and flexibility, all "which impact the ability to manipulate quantity, retrieve facts and resolve intrusion errors." (Menon et al., 2020; NIH, 2022).

Whilst not part of the evaluation for the formal and legal designation of dyscalculia they are very real areas of struggle and must be addressed via interventions and necessary supports. Remember that bell curve I shared earlier, it also means that individuals with dyscalculia are each unique in how their disorder presents itself and the uniqueness of the supports they may need. The individual will fall somewhere on a continuum of mathematical abilities that ranges from severe and possibly permanent difficulties to those with only temporary struggles as the maths are mastered (Grant et al., 2020; Haberstroh & Schulte-Korne, 2016; Kroesbergen et al., 2022; Mattison et al., 2023).

To learn and do math problems I must be able to control, plan, remember, and ignore what is going on around me. I must plan what I will do to solve the problem, focus on what the problem is about and not the rest of the class, and recall if I have done similar problems and then let the math regions of my brain do the problem and recognize whether the answer is logical and therefore correct. This also applies to day-to-day math, planning meals and following recipes, time management for getting to work on time and completing projects, following directions. If I have dyscalculia the connections, the neural pathways between the math and the executive control center are not what they should be and need more time and practice to do what my peers do. And these executive functions are critical to all of the things happening in school and my life. But executive functioning is not more important than the math regions, rather early math development seems to improve executive functioning and enable even more math learning. So, a generative cycle of growth that is critical not only for math but for life.

For the child with dyscalculia these domain-general abilities or cognitive processes manifest themselves as struggles with some of the following.

Working memory: Required for learning and problem solving, working memory is the storage, processing, and recollection of the most recent version of verbal and visuospatial information, in this case the processing of numerical knowledge (Attout & Majerus, 2014; Mahmud et al., 2020). This is the ability to hold in readiness memories of the answer to part one of a multistep problem with numerical information while working on the next part of the problem. Consider the problem, (7 x 3) + 15. When doing mental math, a student would need to multiply 7 x 3 for 21 and hold that number in their head and then add 15. Working memory allows short term storage, but for those with dyscalculia the ability to remember 21 is not always possible. In addition, symbolic information is usually more difficult to access than non-numerical. Remembering 21 is more difficult than remembering that it was about dogs.

Think tennis balls and how many you can hold in one hand. It is probably three to four, and if someone tosses you one more you don't drop just one but rather all of them. And if your dog runs over to play because they see the balls then the probability you will drop all of them increases as your brain determines if he is going to jump on you to invite you to play. That is the impact of the amygdala and the ability to focus on holding the balls rather than being distracted by the dog running at you.

Now think about doing a math problem. A multistep word problem on a high stakes test or a real life on the job problem. I think you are beginning to see the complications dyscalculia can cause.

Attention: The ability to focus on the task at hand rather than other things in the classroom and not be easily distracted is also an area of struggle. Also called inhibition, this is especially apparent when doing multi-step problems and extraneous details need to be ignored or the student has not achieved automaticity for math facts (Kroesbergen et al., 2022, Watson, et. Al., 2016). The student cannot stop themselves from being distracted and thus has difficulty in both the classroom and in daily life. Think of the game 'Simon Says' When you hear "touch your toes" you must take a moment to determine if Simon said you could or not touch your toes and not be distracted by everyone around you starting to make their hands touch or not touch their toes.

Processing speed: Students with dyscalculia also typically require longer periods of time to complete the problem or retrieve the needed information to complete the problem. Processing speed has been identified as a central deficit for students with dyscalculia. (Price & Ansari, 2013). Imagine the class doing a multistep word problem, if I have dyscalculia, ADHD, or dyslexia, while the class is processing step 3 of the problem I am still on the first step. This could be a word problem or a simple 3-digit addition problem, 367+456. This also carries over to multistep directions for completing a project or finding a location.

Phonological processing: Phonological processing is the ability to recognize and manipulate language sounds, and it is a critical skill for developing reading and math proficiency (Bulat et al., 2017). Phonological processing, also known as rapid naming, is the use of the sounds of one's language (i.e., phonemes) to process spoken and written language (Wagner & Torgesen, 1987). The broad category of phonological processing includes phonological awareness, phonological working memory, and phonological retrieval. Hearing and quickly processing the words three and four or accessing the number fact of 3+4 =7 is necessary to do everyday tasks in both arithmetic calculations and word problems.

Visual-Spatial skills: Spatial sense is an understanding of shape, size, position, direction, and movement – being able to describe and classify the physical world we live in (Kroesbergen et al., 2022). Kroesbergen found evidence for the relationship between children's math skills and their visuospatial skills, conservation abilities, and processing speed. He also found that they help explain deficits that are specific to mathematical difficulties. Think about the ability to use a map or recognize that there are columns and rows when working a multidigit math problem or use a graph.

Logical/non-verbal reasoning: The process of solving problems and forming concepts without using words or language is called non-verbal reasoning (Huijsmans et al., 2020; Kroesbergen et al., 2022; Träff et al., 2016). It is used when determining symmetry as in are two shapes equal, are they the same size. Logical reasoning is an interconnected skill, partly comprising conceptual understanding and identifying the next steps to a problem solution (Peters et al., 2020; Träff et al., 2016). Visual perception and ordering, which are used when identifying logical constraints within word problems and if the solutions are logical is also an issue (Duran et al., 2020; Morsanyi et al., 2018; Sasanguie et al., 2017).

Number-specific executive functioning: This is also a key area of logical reasoning. Identifying and categorizing number-specific deep conceptual understanding allows students to see patterns and interrelated concepts. This impacts the ability to remember the last number counted, or to bring down the number when doing an arithmetic problem. It also impacts the ability to remember if an object has already been counted. Consider completing a word problem, the student must determine what operation or operations to do, ignore distracting information in the problem and focus only on numerical information, and then put the operations in the correct sequence.

Math anxiety! Let's add one other area of struggle that is extremely common but is more a result of the other areas not being addressed. That area is math anxiety, and it is definitely not limited to those with dyscalculia. Nearly everyone has experienced the sudden stomach clenching, sweaty palms, lightheadedness, panic, paralysis, and inability to think reaction that is known as mathematics anxiety when faced with a pop quiz or major math test. And usually they are not mild feelings, they are negative, and extreme. Math anxiety is an incredibly important issue to address, especially for students who are struggling and know that they do better in other subjects but for some reason not math.

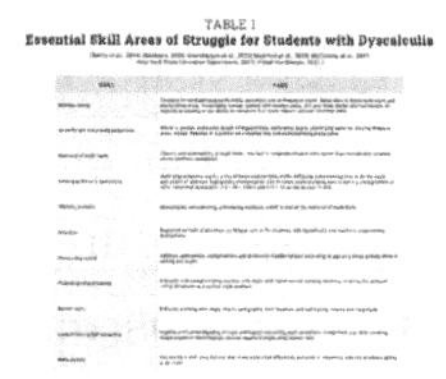
TABLE 1

Essential Skill Areas of Struggle for Students with Dyscalculia

Essential skill areas of struggle for students with dyscalculia

Chapter Three

WHAT CAUSES DYSCALCULIA?

Neuroscience has made strides in identifying the underlying causes of dyscalculia as a neurodevelopmental disorder that impacts reasoning about numbers. Over the years, there have been a variety of theories about the causes of dyscalculia, but currently there is general consensus that it is a multicomponent neurodevelopmental disorder that comprises impaired development and disruptions in the neural pathways in the distributed inter-connected regions of the brain that process numerical information and perform math problem-solving (Räsänen et al, 2021). More recent fMRIs indicate that multiple regions of the brain; frontal, parietal, temporal, visual and hippocampal are activated during the performance of number processing, which also involves recognition and memory processing (Bulthe et al., 2018; Dinkel, 2013). This supports the theory that the disruptions in processing are not limited to strictly math but also more generalized processes such as anxiety, attention, short term memory, etc., and that interventions could be more effective if they address processes other than math to support the individual with dyscalculia. This also supports the growing consensus that dyscalculia exists on a continuum and individuals will exhibit a variety of issues not just math and is as unique in its profile as the folds of the brain are unique to everyone. Even identical twins do not have identical fingerprints or folds to their brains (Dehaene, 2020).

Although research into dyscalculia is still not as common as it is for the related learning disorder, dyslexia, there is a growing body of evidence and acceptance that it is due to genetic risk factors rather than environmental and occurs during the early stages of the brain's development (Dehaene, van Bergen, et al., 2023). If one were to describe the impact of genetics on dyscalculia it would be to describe how genetics influences the development of our brain, the unique folds of the brain that happen in utero and become the various regions. Some regions are specific to math processing (domain specific) causing deficits in number processing, number sense, etc. The other regions (domain general) cause the more generalized deficits in executive functions such as working memory, phonological processing, visual spatial memory, attention, etc. In turn, we can see these manifest as difficulties in learning number sense, math facts, math calculations, mathematical reasoning and in performing math in day-to-day life.

Regions of the Brain and their Roles

Why is it important to understand the role of the regions and the brain in working with students with dyscalculia? Because math is not dependent upon one area of the brain but rather the ability to connect between a number of regions, all of which play an integral role in successfully using math. When we learn and do math we build a mental image of the objects in the problem, we then use the math regions of the brain to do the problem and it is our executive region of the brain that tunes out the child in the corner singing, our need to ask the teacher to go the restroom, and in the end enable us to solve the problem and get the answer.

It is easy to think that learning the concept of a number such as 3 is easy. The typical child grasps and uses the numeral so quickly that it appears that it is as simple as connecting the sound of the word t-h-r-e-e with 3 objects, but the child also has to write the symbol 3 or write the word three and understand that 3 is tied to a quantity and not a single object. All of those must be tightly connected to form the concept of "threeness". This involves connection and communication across the following regions of the brain – all of which play a key role in mathematics.

And with dyscalculia, advances in our understanding of neuroplasticity support the idea that the brain and its neural connections physically change in response to teaching and improve the ability to hold memories in long term memory and suppress incorrect memories (bin Ibrahim, 2021; Dehaene, 2020). How long these physical changes last is one of the most important areas of research that needs to be done. Being able to predict how long one can remember the math facts, the rules, will help us to determine how to space practice for longer recall. With early intervention and the right supports, including accommodations and modifications, it is very possible to adapt to being dyscalculic and demonstrate mastery of mathematics and be successful as a student and adult (Bailey et al.; 2020).

Three areas of the brain need to communicate in order to answer a question about which number is larger, 3 or 5. These are the:

- parietal region housing the IPS (intraparietal sulcus), where the magnitude or which is larger 'concept' resides,

- the frontal region housing the prefrontal cortex which makes sure the student pays attention,

- and the hippocampus, which will process the concept of which is larger, 3 or 5, for storage in long term memory (Al Otaiba & Petscher, 2020).

They are all connected through white matter, nerve fibers that connect the 3 regions of the brain and which determine the ease or speed with which an individual can later retrieve the information needed to successfully work the problem (Nicolson & Fawcett, 2021; Watson & Gable, 2016, Vandecruys, et al., 2024). The larger the neural connections, the nerve bundles, which are a result of practice will determine how quickly and accurately the student can answer the question of which number is larger, 3 or 5.

Frontal Region – The frontal region of the brain is responsible for executive functions, decision-making, reasoning, personality expression, social appropriateness, and other complex cognitive behaviors. Prescott et al. (2010) found that the connectivity patterns in the frontal region are consistent with previous studies linking increased activation of the frontal and parietal regions with high fluid intelligence and are possibly a unique neural characteristic of the mathematically inclined areas of the brain (Padmanabhan & Schwartz, 2017; Siemann & Petermann, 2018). The frontal region is directly linked to conceptual understanding and mathematical reasoning. This may explain why students struggling with dyscalculia show persistent deficits in number processing, which is associated with connectivity of the frontal and parietal regions of the brain (Räsänen et al., 2014; Soares et al., 2018; Van Viersen, 2013).

The prefrontal cortex is the very front of the frontal region and houses the executive functions that enable us to problem solve and pay attention, which is necessary when learning how to map the heretofore meaningless numerals and words to their appropriate magnitude meaning when acquiring number sense or focusing on the operational sign to do a single or multi digit computation problem. The ability to plan and sequence a multi-step problem, ignore distractions or superfluous information in a word problem, and stay focused for the extended period needed to solve the problem are also tasks that the student with dyscalculia struggles with (Prescott, et al., 2010). Attention to task is also correlated with better counting, numerical processing, and calculation skills (Lynn, et al. 2022).

Parietal Region – The parietal region is the primary sensory area of the brain and is responsible for sensory processing. A specific area of the parietal region, the IPS or intraparietal sulcus, allows us to discriminate between quantities, a key component of number sense (Price & Ansari, 2013, Ustun et al., 2021). Research suggests that individuals with dyscalculia, as compared to students who are simply struggling with math, have difficulty differentiating between a set of 5 and a set of 6 (Mazzocco, et. al., 2011). If they have difficulty comparing the magnitude of the symbols, knowing that 6 is greater than 5, there will be difficulty with math achievement (Bugden et al., 2014).

The intra-parietal sulcus (IPS) plays a major role in arithmetic tasks and more complex mathematical tasks and fMRIs have shown reduced neural activity in the IPS during the performance of assigned tasks for individuals with dyscalculia (Kucian & von Aster, 2015). Depending upon the researcher, this deficit, disorder, difference, weak or slow development of the neural pathways or the white matter connecting these areas; can lead to decreased efficiency or ease of access to the areas of the brain involved in mathematical processing, specifically for arithmetic and numerical problem-solving. This underdeveloped interconnectivity is a risk factor for dyscalculia (Kuhl et al., 2021; Martin & Fuchs, 2022; Menon et al., 2021; Grant et al., 2020; Price & Ansari, 2013, Ustun, 2021).

Hippocampal Region – The hippocampus is a complex brain structure embedded deep in the temporal lobe and has a major role in learning and memory. According to Üstün et al. (2021), the hippocampal region of the brain is activated during symbolic number perception, particularly in students with dyscalculia. The hippocampal region is tied to operations and algebraic operations through the memorization of arithmetic facts and accurate mathematical reasoning. Interestingly memory is strongly tied to sleep as it is during sleep that memories and learnings from the day are consolidated and stored for future retrieval. Quality sleep has been found to be extremely important during childhood as foundational math information is being learned and it is during sleep that the hippocampus stores the information of the day in long term memory (Hoedlmoser et al., 2022).

Two other regions of the brain are also involved as one must hear the math problems and the instruction, and one must also see. These regions are the temporal and visual regions.

Temporal Region – The temporal region is associated with processing auditory information and with the encoding of memory. The temporal region of the brain is most directly related to elementary number sense and memory related gains or deficits (Kroesbergen, et al., 2022). Anobile et al. (2022), found that dyscalculia and neurotypical development suggest that visual perception of numerosity (number sense) is a building block for math learning and is directly linked to the encoding of memory for mathematics.

Visual Region – The visual cortex is the primary cortical region of the brain that receives, integrates, and processes visual information relayed from the retinas. According to Bulthe et al. (2019), it is responsible for processing visual information, including numerical symbols, which are important for mathematical reasoning. In the classroom, the visual cortex is closely related to accurate mathematical reasoning and of course the visual input of information needed to do the math.

Individuals with dyscalculia show difficulty in using symbolic (Arabic numerals and number words) and non-symbolic representations (dots, sticks, 2-dimensional and 3-dimensional items) when asked to do tasks that involve represent, access, or manipulate number sense (Geary et al., 2013; Mazzocco et al., 2011). Mathematics is built upon the ability of our brains to map a symbolic number system (Arabic numerals and words) on top of the pre-existing visuospatial number system (the visuals we store in our brains of sets of twos and fours, both 3 and 2-dimensional). This includes subitizing, attaching threes to the abstract representation 3 and discriminating between larger numerosity's without counting - recognizing the five hearts on a playing card without counting, grabbing three cookies without having to count them out. These differentiation skills can be observed as early as 6-months of age and continue to develop, so they are early indicators of the potential for dyscalculia. In addition, research is suggesting that counting (the ability to orally sequence numbers correctly) is also a good predictor (Geary, 2013).

If you were to look at the brain using an fMRI, the areas that show increased oxygen use or activation during a math task are in the areas of the brain discussed above. Dyscalculic students performing a non-symbolic numerical comparison showed no activation of the right IPS and atypical structural development of the IPS compared to students with typical development. This is now believed to be the slower formation of white matter or myelin sheathing of the neurons which translates to slower speeds of being able to access and process problem-solving requests by the brain (Dehaene, 2020; Fields, 2014; Price & Ansari, 2013). Research by Kuhl et al., (2021), expanded on this and found that they could predict with 87% accuracy, students who developed dyscalculia, based upon the functional connectivity of the right posterior parietal cortex (PPC) and the right dorsolateral prefrontal cortex (DLPFC) and the effective connectivity of these two regions by white matter, the neural connections. These are the areas of the brain that are most strongly associated with numerosity and calculation, respectively, and the poor structural and functional connectivity supports the explanation for why people with dyscalculia need more time to solve problems than people without dyscalculia.

This combination of difficulties and strengths we believe provides a more accurate picture of the individual at any point in time and allows us to determine when to intervene if needed. Math performance exists on a continuum across the general population and within the individual. An individual with dyscalculia may have neurological issues that cause deficits or inefficiencies in working memory, engagement, number sense and math facts, all of which are highly correlated with success in math learning. Poor performance in math in comparison to normal performance in reading or other areas can cause feelings of stupidity, failure and 'imposter syndrome' and other impacts on mental health. Dyscalculia is impacted positively and negatively by a wide variety of factors with the neurological ability to successfully learn mathematics being multiplied, exacerbated or mitigated by environmental and biological factors. These include environmental factors such as where you live, socioeconomic variables, attendance, adult expectations, trauma, etc. There are also biological factors such as intellectual abilities, emotional factors and neurological variables (Butterworth, 2003). All the above have a direct impact on an individual's ability to successfully learn mathematics.

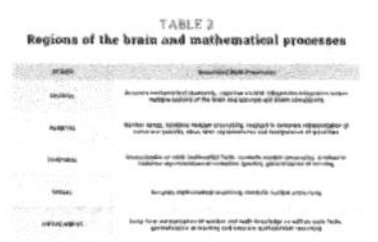

TABLE 2

Regions of the brain and mathematical processes

Regions of the brain and mathematical processes

Chapter Four

BUT WHAT DOES IT LOOK LIKE?

If we think of our abilities across so many things, to compete in the Olympics, to make an incredible street taco, to parallel park, the profile of who we are across our abilities compared to everyone else can best be explained by the simple bell curve. In the area of mathematics, some of us can excel at statistics and string theory, others at the less complex math needed for accounting, to those for whom math takes more time and those for whom cognitive issues interfere with the ability to perform simple math or recognize numbers and their associated quantity. Remember the bell curve I shared earlier which illustrates where we could each perform in relation to others.

If I am a parent or a teacher, what do I see my child doing that may tell me this is more than just a typical temporary struggle?

The PreSchooler

- Has trouble learning to count and skips over numbers long after same age peers remember the numbers in order; 1,2,3,4,5,6

- Doesn't seem to understand the meaning of counting- for example, when you ask for 5 blocks, your child hands you a large group of blocks, rather than counting them out. In the case of 2 or 3 blocks, they should 'know' what 2 or 3 blocks are without having to count them out to the adult.

- Struggles to recognize patterns, like smallest to largest or tallest to shortest
- Has trouble understanding number symbols and math vocabulary, like making the connection between the numeral 7 and the word seven
- Struggles to connect a number to an object, like knowing that the numeral 3 applies to groups of things like three cookies, or three friends

The Elementary School Child

- Has trouble learning and recalling basic math facts, like 2 + 4 = 6
- Still uses their fingers to count instead of using more advanced strategies (like mental math)
- Struggles to identify math signs like + and – and to use them correctly
- Has a tough time understanding math phrases, like greater than and less than
- Has trouble with place value, often putting numbers in the wrong column
- Slow to perform calculations compared to their siblings or other children
- Weak mental arithmetic skills, adding numbers in their head results in more mistakes and frustration than other students.
- Addition is often the default operation. When asked what 4 x 4 is they will add 4+4+4+4 to reach the answer as the quick, automatic recall of basic math facts has yet to be mastered
- Cannot remember their parents' phone numbers

The Middle and High Schooler

- Struggles with math concepts like commutativity (incorrectly believing 3 + 5 is not the inverse of 5 + 3,) and inversion (being able to solve 3 + 26 – 26 without calculating)
- Has a tough time understanding math language and coming up with a plan to solve a math problem
- Has trouble keeping score in sports games and gym activities
- Has a hard time figuring out the total cost of things and keeping track of money (such as their lunch account)
- Avoids situations that require understanding numbers, following multistep rules or procedures, this includes games that involve math. or for which the rules do not seem to make sense.
- Exhibits extreme math anxiety, avoids going to school on test days, skips math class more than other classes, is 'sick' or has excuses for not doing math classwork or homework
- Struggles to read charts and graphs
- Has trouble applying math concepts to money, like making exact change and figuring out a tip
- Has trouble measuring things like ingredients in a recipe or liquids in a bottle
- Lacks confidence in activities that require understanding speed, distance, and directions, and may get lost easily
- Has trouble finding different approaches to the same math problem, like adding the length and width of a rectangle and doubling the answer to solve for the perimeter (rather than adding all the sides)

The Adult

- Remembering numbers to transcribe them - think 2 factor authentication
- Entering the correct PIN to do a bank withdrawal
- Avoiding financial decision due to a lack of understanding - bill payments late, credit cards maxed out and bank accounts overdrawn
- In a longer math problem, remembering the earlier answers in order to find the later answer
- Has difficulty being on time to work and meetings or paying bills on time
- Gets lost easily, remembering and following directions to find a location are a struggle
- Difficulty transcribing the prescription numbers into the phone to refill a prescription

Comorbidity with other disorders such as ADHD and dyslexia and why it is important

Dyscalculia can occur as the primary disability-causing problem with learning math, or it can occur as a secondary condition, comorbid with ADHD, dyslexia, dysgraphia, and reading comprehension disorder. It has been relatively common for research to report comorbidity of specific learning disorders, usually dyslexia with other disorders and ADHD with figures swinging from 35% to 80% but recent research (Dewey and Bernier, 2016; Nuffield Foundation, 2018; van Bergen, et al., 2023) supports that the shared genetic risk factors present in the population of those with ADHD, dyslexia, and dyscalculia predict the cooccurrence of each other but not the guarantee that an individual will have all 3 disorders. Their study of approximately 19,000 twins and 2,150 siblings yielded 77.3% of the children having only one disorder with 22.7% having one or more. A review of the research by Aksoy (2024) yielded a comorbidity rate of 31% to 45% between learning disabilities and ADHD.

Because research around dyscalculia alone has lagged and as a result it is less well known, and there are fewer screening and diagnostic tools (Jaya, 2009), but its prevalence is estimated to be similar to dyslexia - 5-7% of the population (Menon et al., 2020; Price & Ansari, 2013; Santos et al., 2022). If we use the population of the United States in 2023, 335 million individuals, that would calculate to over 16 million individuals having dyscalculia severely enough to impact daily life and career activities. If we consider the 2022 PK-grade 12 public school population of 50 million students that will yield approximately 2.5 million students if we use the conservative estimate of 5%. Or nearly the entire population of the state of Kansas. Add to that the research around dyslexia and the estimated comorbidity of 35-70% with other neurological disabilities appears realistic. That is a staggering number of students who are at risk of failing math when early identification and intervention can mitigate the problem in many cases (Kisler et al., 2021; Litkowski et al., 2020).

A note of caution is warranted here, even though the prevalence and comorbidity of dyscalculia is similar to dyslexia, treating one will not necessarily improve the other or others as the case may warrant. What research has shown is that students with dyscalculia have higher rates of ADHD and that even when other disabilities are addressed there is still a higher occurrence of anxiety specifically around math and tests (Grigorenko et al.; 2020: Willcutt et al., 2013).

The child who is diagnosed with the more commonly screened for ADHD or dyslexia may or may not have dyscalculia, but it is highly likely that the child will struggle with math due to the shared areas of the brain that are impacted, and dyscalculia is a possibility and requires specific interventions for the math issues. In turn, there may be no dyscalculia and only a diagnosis can determine what is needed.

Weaknesses in mathematical reasoning, phonemic awareness and engagement result in poor comprehension of word problems and the deficits in visuospatial memory contribute to the difficulty in recognizing and mastering math facts. So, a child with ADHD may show the same difficulty as a student with dyscalculia but not have both disabilities, stressing the need for accurate diagnosis and intervention as the needs of the various groups are similar but not always the same. Examples of commonly shared difficulties across dyscalculia, dyslexia, and ADHD include word problems, mental math, slow processing speed, and response time to questions, completing multi-step problems, remembering sequences of numbers or patterns, working memory, telling time, retrieving facts, etc.

In Summary

There appears to be a general consensus that genetic factors and the resulting disruptions and inefficiencies in the specific neural pathways responsible for math and those responsible for more generalized process such as working memory, visuospatial memory, phonological processing, attention, executive functioning are the major causes of dyscalculia. This results in a variety of weaknesses in math pertaining to number sense, successfully learning the math facts, performing calculations and the area of mathematical reasoning. But we know that current research strongly supports that with the appropriate interventions from parents and educators, the impact of dyscalculia can be mitigated.

Part II

The goals of this chapter are to help the reader:

- Understand why screening is important and should be done as early as possible.
- Understand some of the screeners and diagnostic instruments that are available.
- Understand the information a screener can provide the diagnostician and how it can be used to immediately target the areas of struggle regardless of the outcome.

Chapter Five

THE IMPORTANCE OF SCREENING EARLY

What is a Dyscalculia Screener?

According to experts, the purpose of a screener is to sort students quickly and accurately into groups of yes or no to determine if further evaluation is warranted, in this case for dyscalculia. Sometimes, the screener can include maybe as a designation, indicating that there are some risk factors. Screeners should identify children at risk and minimize the chances of inaccurately identifying a student (Grant et al., 2017), and they should allow for the maximum amount of information in the minimum amount of time. They should indicate where the potential for dyscalculia exists and provide the impetus for a diagnostic process to begin that will formally identify the individual has dyscalculia.

In the case of dyscalculia, the screening is for evidence of the indicators the DSM-5 requires being used to label an individual legally as being dyscalculic. These are number sense, mathematical facts, math calculation, and mathematical reasoning, the early math skills. These foundational math skills are predictive of later math achievement, which makes a screener useful for instructional decision-making purposes, regardless of whether a diagnostic process is completed. A screener should be easy to find and easy to use for those that wish to learn more about why a student is struggling with math and how to best address the issue. It should also be low cost or no cost so that money is never an impediment to providing services to those that need them.

In summary, a screener 'shortlists' those who will need further testing and possible 'diagnoses, if they are unsuccessful in mastering specific arithmetic and problem-solving domains and enables the educator to provide appropriate interventions and supports in addition to data to determine if a diagnostic process needs to be started. It cannot be used for diagnosis as it is a quick assessment rather than an in-depth evaluation.

Because a screener can identify individuals who are struggling with mathematics and can be used to determine if a diagnosis process should be started, screening should happen as soon as a child has significant problems with the early foundations of math development, namely number sense, memorizing math facts, accurate calculation skills, and mathematical reasoning. Early identification and intervention can prevent further academic struggles not only in school but throughout adult life as research strongly supports the positive impacts of intervention to the point of being able to do as well in math as their typically developing peers (Abd Halim, 2018; APA, 2022; Bailey, et al., 2020; Chodura et al., 2015; Dennis et al., 2016; Mahmud, 2020; Menon et al., 2021; Mononen, 2014).

Struggles with math during the school years and even through adulthood that warrant screening are indicated by assessment scores lower than the 25th percentile on standardized formative and summative tests (Kong et al., 2022; Swanson et al., 2013). However, for younger students and children, those below 3rd grade, it can be based on observation of the student by educators and guardians (Mueller et al., 2012).

The screening process should also provide general information about a child, as the struggle with math may be due to factors unrelated to dyscalculia, such as vision, hearing, and language. These should be ruled out before considering learning disabilities or other factors as the underlying cause of the math difficulties (Hayes et al. 2018).

A review of the literature as well as internet searches indicate that there are several screeners available that assess some or all these skills and they come in varying price points and languages. Some follow the required areas laid out in the DSM-5 and others do not. Personally, I feel that it is important to choose one that provides the individuals who are responsible for diagnosing dyscalculia, the diagnosticians and psychologists, as much information as possible to select the appropriate diagnostic instruments and not over test the child.

Why Screen?

Why screen? Why not assess and move to diagnosis as fast as possible? My response would be that there are too many students to try and have all of them go through a formal diagnostic process when current observational and assessment data available with a screener will quickly sort the students who are most likely to qualify for further assessment and potentially diagnosis. Time is of the essence, as early identification of dyscalculia and others who struggle with math coupled with appropriate interventions is more effective and less costly than waiting (Nelson, et al., 2022). With intervention there is less probability the student will fall behind and typically once they fall behind, without intervention, they will stay behind. In addition, with intervention, it is possible for those with dyscalculia to perform math as well as their typically developing peers.

As stated earlier, to be diagnosed as dyscalculic, the DSM-5 requires impairments in mathematical skills such as number sense, memorizing math facts, math calculations, math reasoning, and math problem solving (APA, 2022). These are the areas of early math performance that research supports as being commonly found in individuals who are later diagnosed as having dyscalculia as well as individuals evidencing lower-than-expected math achievement (Gersten, 2011; Jordan, 2013). The use of a dyscalculia screener allows these areas of risk to be identified early in a child's life through tasks that measure those specific skills. With the increased usage of universal screeners for reading and math as well as beginning of year placement tests that start as early as PK, it should be possible to screen as early as PK for dyscalculia by examining the student assessment data for those who scored below the 40th percentile. Most of the students who will later be formally diagnosed with dyscalculia will typically be those in the 15th percentile and below as well as other disabilities that can interfere with math learning. This would enable interventions to start at a time when brain plasticity is at a higher level as it gradually tapers off throughout our lives. The potential for an extra 5 years of intervention at an even more impactful time is an important goal. Waiting until the end of third grade to review data from mandated state assessments is a loss of precious intervention opportunities.

Who should be screened?

Who should be formally screened for dyscalculia or the potential for math difficulties? Shouldn't every child? But what about the current assessments that seem to be ongoing once a child starts in school, whether it be as a 3-year-old in the early learning programs or as kindergartner, what about the tests that happen every fall, winter and spring starting in many 1st and 2nd grade classrooms or the state assessments that happen to all 3rd grade and up students?

Before formal screening happens, there are often informal assessments and observations that have been done that can be used to initiate the more formal process of administering a screener and/or diagnostic instruments. This information provides important insights into what might be causing or contributing to the student's struggles. Observation happens anytime a parent or educator notes something that is out of the ordinary when a child is performing an activity involving math.

At home or in a PreK class, you can notice that a 3-year-old cannot distinguish between a larger and smaller dog, cannot hand them three cookies, takes a long time to respond to the requests or does not seem to grasp the concept when shown something - there are more blocks in this bucket than that. When shown something similar it seems they do not understand the concept even with repeated explanations. An elementary teacher would also notice slower processing times, avoidance of math activities and games, failure to master all of the arithmetic facts, or knowing them one day and not the next. A common one is continued use of finger counting or difficulty explaining how they solved a problem or an illogical reasoning process. These observations are extremely important in rounding out the information on the child, their more domain general functioning and environmental factors such as the completion of a vision screening or high absenteeism as math struggles could be caused by other factors, not dyscalculia. A key indicator is that the child's actions are not those expected for the typical child when doing math.

Informal observations or the use of checklists are easy ways to gather basic information that can help decision makers decide on next steps such as administering a vision or hearing screening, providing additional instruction because the student has missed 3 weeks of schools, etc.

Chapter Six

WHERE CAN I FIND A SCREENER FOR DYSCALCULIA AND IMPORTANT COMPONENTS?

Compared to the number of screeners available for dyslexia there are few for dyscalculia but as dyscalculia becomes a larger part of the discussion around why we have students who struggle with math the numbers will increase. In selecting a screener to use, important factors that are useful are speed of administration, collection of information on math performance for number sense, arithmetic facts, mathematical calculations, and mathematical reasoning, the ability to be administered in a group or individual setting, collection of observational data around the student or individuals daily math practices and how they approach math. In the US, these are the key areas that diagnosticians, school psychologists and those who can legally provide the diagnosis of Specific Learning Disability-Math or dyscalculia use to make the determination.

The collection of observational data in a survey form is usually overlooked during the screening process but is especially important as it rounds out the data that is collected and may point to other causes of the math struggles. Information that should be part of a screening includes whether the child has had a vision or hearing screening, has there been a diagnosis of other exceptionalities, is another language spoken in the home, has the child been frequently absent from school, etc. All of these could contribute to the difficulties the child is experiencing and need to be addressed. In the case of older children, they may have found ways to accommodate much of their disability and the results from the screening portion of the process would not indicate any risk factors but behavioral factors such as avoidance of math activities, math anxiety, the use of fingers to count with long after their peers have stopped, knowing their math facts one day but not being able to recall them or being very slow the next day are not discoverable during a timed screening process. Thus, the need for a survey portion in the process of collecting data.

Here is what I have found to date and what we have developed to meet what we believe would be helpful. I continue to monitor the research for additional screeners, with awareness of dyscalculia increasing there will be a need for more and I am surmising that it will much like the process was for dyslexia. Many of the screeners I have found to date are from countries that belong to the European Union or are from the United Kingdom and are useful to provide you with a broader view of how the rest of the world views dyscalculia. I have also included the screener that I led development for at TouchMath.

Here is another note of caution from me, I am in no way endorsing any of these. You will want and need to review them and decide which one bests suits your needs. To help speed up that process and help the students who are struggling, I have shared what I know to date so you can do your due diligence.

Two of the current screeners are the Preschool Early Numeracy Scales (PENS; Purpura & Lonigan, 2015), and the Number Sense Screener (NSS; Kaufmann et al., 2013). Both assess the numerical skills of counting, one-to-one correspondence, number-word-knowledge, etc., and are targeted at preschool to first-grade children and take approximately 10 minutes to administer. Other screeners include the Test of Early Numeracy Curriculum-Based Measurement, developed for screening the younger child and a similar paper and pencil screener, commonly referred to as the "Brief Assessment of Number and Arithmetic Skills" or BANA. It was released in 2008, but accuracy rates were only at 51% (Bugden, et al., 2020).

Butterworth (2003) created the Dyscalculia Screener which focuses on tests of basic numerical capacities and numerosity. The concentration of questions encapsulates numerosity as a property of sets, estimated numerosity, magnitude, and basic counting, number Stroop, dot counting, item-timed arithmetic, and simple reaction time. It is widely used in the UK, has a cost, and is the foundation for many, if not most, screeners as they are based on his early and continued research on dyscalculia.

The Numeracy Screener (Nosworthy et al., 2013) is a 2-minute test of symbolic (Arabic numerals) and non-symbolic (dot arrays) discrimination ability developed in 2013. While it predicted a large percentage of children who evidenced signs of dyscalculia, it was focused on only two areas of math ability. It was found to have only poor to fair levels of clinical and practical significance (Bugden et al., 2020). It was recommended that additional tools be used to supplement it for screening for dyscalculia. It was also recommended that screening tools that measure symbolic abilities would be better for use as a screener.

Gliga and Gliga's (2012) screening instrument looked at the sub-categories of estimating quantity, counting backward, or repeated sequence counting. This screener's theoretical focus is on the Triple Code Theory or TCT, which proposes the existence of independent number mental math representations.

The Dynamo Assessment was created in 2016 to address deficits in students with mathematical difficulties. The assessment has 647 test items and is used to screen and create a general intervention and support plan focusing on students ages 6+. It was normed in the United Kingdom and has a cost associated with its administration as its intent is to provide the intervention activities.

Rasanen et al. (2021) released a study on the development of a Finnish project to develop the Functional Numeracy Assessment. A subproject was the development of a screener for dyscalculia. It used six tasks to screen for the disability- number comparison, digit dot matching, number series, single-digit addition, single-digit subtraction, and multi-digit addition and subtraction. They were able to show that these tasks could be used across genders and from grades 3 to 9. To date, it is available only in Finnish.

There is also the Dyscalculia Screener (Wells, 1997) by Schreuder. This instrument includes 15 modules, encompassing the approximate number systems, focusing on grades PK to 9th grade. A small survey is connected to the screener. Its goal is to provide specific grade level math information that can be used by therapists to build and execute an intervention plan.

Grigore (2020) developed the Diagnostic Assessment of Dyscalculia for diagnosing and identifying dyscalculia and is focused on basic number processing skills. The components of this tool include basic processing skills such as enumeration, linking non-symbolic representations to symbols, transcoding, counting, number comparison, and measurement (number line and analog clock reading). While it is more in-depth than Butterworth's (2003), to date, it is only available in Turkish.

In 2023, it became apparent that there was not a screener developed for the US population that I felt addressed the criteria that diagnosticians had to use so I assembled a team, and we developed the DySc dyscalculia screener. It is based on previously proven question types for assessing the four areas diagnosticians need to examine. It can be administered individually or in a group setting. Data collection and further refinements will be an ongoing process as predictive validity and other measures are conducted in addition to its current content validity.

It includes observational information from educators and guardians that may more easily and accurately explain a child's math struggles, namely the failure to pass a vision or hearing screening or having a different primary language at home than is used in school to teach math. In addition, it can screen various age groups starting as young as 3, as early intervention has been shown to improve math performance long term. It can be administered with minimal preparation and no financial investment, via print materials or digitally, covers ages 3- adult, and is available in English and Spanish.

The DySc also provides what other screeners do not. The report it generates includes an action plan and recommended evidence-based interventions to implement while diagnostic procedures are ongoing. You can access the DySc Screener for Dyscalculia at dysctest.com.

Chapter Seven

YOU'VE SCREENED – NOW WHAT?

After receiving the information from a screener, it is critical to act because the earlier the intervention the better the chances of decreasing the impact of the issue with math even if dyscalculia is not the end result.

If the screener returns results of:

NO – there are no indicators of risk factors for dyscalculia during this screening. It is possible that they are not apparent at this time, so it is important to continue monitoring math progress and if there is still an issue then retaking the screener is a reasonable action. Screeners should provide information on how often they can be retaken before the results are due more to the student being familiar with the questions than to there being indicators of dyscalculia. At the same time, there is still the issue of why the child is struggling with mathematics and the answer needs to be found. But while the process to determine what the cause is, it is imperative that math struggles be addressed through intervention and support in the areas that are a struggle and to assure the student that the adults in their life are working together to do something about it.

MAYBE or SOME – if the screener returns evidence of some risk factors, then while intervention and support is provided to the child, there are several paths that can be taken. Parents always have the option based upon affordability to seek an outside assessment and diagnosis from professionals such as clinical psychologists and licensed diagnosticians. Please ask your school and district administrators, pediatrician, or your state or local chapter of advocates for exceptional children or learning disabilities for suggestions on next steps.

It is also reasonable to immediately provide math interventions in the area of struggle and continue to collect data via a screener three to four months later. Continued monitoring and data collection could show that the child still struggles, and diagnostic assessment is warranted to uncover the cause, or the child improves and no further action other than monitoring is warranted. In most schools, Multi-Tiered Systems of Supports (MTSS), is in place and is a viable and proven effective option to provide intervention and ongoing monitoring so as not to lose time.

YES - when a screener indicates that there are definite risk indicators for dyscalculia then you are able to request that a formal assessment and diagnostic process begin. The school and district usually work with the parent to move this forward and in those cases where they refuse to, please seek advice from the local and state advocates for exceptional children or the local learning disabilities chapter.

There is also the option of an outside assessment and diagnosis and the use of private providers of educational therapy and private schools. You can find information from organizations such as Learning Disabilities Association of America, the Educational Therapists Association of America, Understood.org, etc. (See the References for a comprehensive list of research and resources)

What to expect from a screener or diagnostic report

Once the screener has been administered and observational data collected via an interview or checklist completion by the parent or guardian or an adult most closely connected to the student, the results will arrive, and it will be time to talk and act. What information should you expect to see and what should the discussion be about that will help drive the best decision and plan for the child? Everything provided in the reports should help inform the decision that needs to be made;

- should we move to a formal diagnostic assessment and a possible legal diagnosis of dyscalculia,
- should we screen again later as there are some indicators but nothing definitive, or
- should we look for other causes of the math struggle.

Regardless of which of the three decisions is chosen, interventions to help the child should begin immediately. Screening reports will vary based on the screener, but there are key pieces that should be included.

Reports such as these should be written in a clear and easily understood language with as little jargon as possible and clear explanation of terms such as 'dyscalculia', 'learning disability', and acronyms such as SPED, ADA, IDEA, etc. Examples and non-examples are always helpful in explanations. Because so many individuals; educators, parents, and the students themselves will be reading the information, and it should be clear that all the information was accurately captured especially where an interview is part of the screening. During a screening conversation there will be a wide variety of data that should be shared, ranging across classroom worksheets, homework, observational surveys, state-level, and standardized formative assessments. These provide additional information about how the student is progressing. Formatting the report so that data is easily read and interpreted should also be expected. Visualizations of data such as graphs that compare current to expected, the use of colors, etc. help to make information more easily digested and understood. These reports should also include the next steps or actions to take; additional assessments needed, what interventions can be undertaken immediately, and the specific areas of weaknesses. The inclusion of resources that the parent and student could find informative would also be useful.

It is reasonable to ask if everything has been considered when reviewing the report. That is 'job number one' that you have as a parent or teacher.

Moving to assessment and diagnosis

After the results of the screener are reviewed it is important to talk to an expert before making the decision to proceed with further formal assessments that can be used for a diagnosis. This part of the process is driven by each state's individual statutes governing who is eligible for the designation of dyscalculia. Your local school district Exceptional Child division or the Child Study Team can explain what is possible in your state.

Since a screener is typically designed to be administered in a short period of time - less than 20 minutes and do a quick sort of students into yes, no, and maybe groups of children about whether or not there are indicators of risk - it is not the definitive tool for deciding if an individual has dyscalculia. Regardless of the results, the screener was probably used because there is a level of concern about the students' progress in math that warranted the gathering of additional data and deciding if there is a problem. So, once the results are available, in the case of a parent, you need to talk to a professional.

If your child is three or under talk with your doctor and they can give you the contact info for your state's Early Intervention program. If your child is older than 3, contact your school district or talk with the teacher, principal or guidance counselor, and ask about the next steps and what is involved. Every school district has a department dedicated to supporting families and students with exceptionalities. They can provide you with more information about the process that school districts must follow as the determination of exceptionality, disability, 504, or IEP is a legal process not a medical one. If you are a teacher, the same advice goes, move it up the ladder to those who can start the process. If you are the parent, and the institution chooses not to proceed you always have the right of appeal and to go outside the system and get a private diagnostic process going.

This is one of the most important decisions a parent can make as the legal decision maker until a child is legally an adult - whether to move to the diagnostic process and all that entails. Diagnosis takes time and can be stressful as there is no way that the student can believe that this is a normal procedure. It can be stressful for the adults if they disagree with the school and its decision to proceed or not proceed, it can cost money, and the outcome can provide answers that help or send everyone looking for another reason. If you're already an adult it can still be stressful, but it usually provides long looked for reasons as to why math has been a struggle in contrast to other subject areas.

It is important at this point to stress the difference between a screener and a diagnostic tool. Diagnostic instruments in contrast to screeners, can only be used by licensed professionals during the process of determining if a disability exists and if the individual is entitled to legal and educational services.

Other forms of educational testing exist but are not designed for quick screening but rather for use in the formal diagnosis of various disabilities such as dyslexia, cognitive impairments, ADHD, etc. The following is not a comprehensive list but some of the diagnostic instruments that are commonly used when dyscalculia is suspected include:

The Wechsler Intelligence Scale for Children, the WISC-V, which measures children's intellectual ability from 6 to 16 years. The WISC was developed to provide an overall measure of general cognitive ability and also measures of intellectual functioning in Verbal Comprehension (VC), Perceptual Reasoning (PR), Working Memory (WM), and Processing Speed (PS) (NIH, 2016).

Diagnosticians are also using the FAM (Feifer Assessment of Mathematics) which is designed for prekindergarten through college students (ages 4-21 years) and consists of 19 subtests. Each sub-test maps onto three domain indexes (procedural, verbal, and semantic) in order to provide specific information that aligns to Feifer's view that there are subcategories of dyscalculia. (Mulchay, et al., 2022).

Another common educational assessment, the Woodcock-Johnson IV Tests of Cognitive Abilities (WJ-IV COG), includes 20 tests measuring four broad academic domains: reading, written language, mathematics, semantics, and theoretical knowledge (NIH, 2021). It includes 18 tests for measuring general intellectual ability, broad and narrow cognitive abilities, academic domain-specific aptitudes, and related aspects of cognitive functioning, which licensed professionals use to diagnose a learning disability, including dyslexia and dyscalculia (NIH, 2021).

In Summary

This chapter has shared the purpose of early screening and that it is critical based on the research telling us that early intervention works. Screeners help to quickly determine if further steps should be taken and if there might be other explanations for the mathematical struggles. Because dyscalculia is a heterogeneous condition the results that come back from a screener will vary from individual to individual and the provision of additional information from observational data helps to determine the next steps. We have also covered examples of the screeners that are available as well as some of the diagnostic assessments that are commonly used and what should be covered in the reports and conversations that are generated by these instruments.

PART III

The Goals of this chapter are to have the reader:

- Hear how adults with dyscalculia successfully live with their disability.
- Hear their advice to those who are supporting children with dyscalculia.

Chapter Eight

LIVING AND THRIVING WITH DYSCALCULIA: A SERIES OF INFORMATIVE INTERVIEWS WITH ADULTS WHO HAVE DYSCALCULIA

During 2024, I conducted a series of interviews with adults who have been diagnosed with dyscalculia. Interviews with adults about what it was like to learn they had dyscalculia, how it impacted their career choices, the adjustments they make to their lives because of the disability, what they would have told their 14 year old selves, and what they would like the current generation of parents and educators to know that might inform what we do with the current and future generations that could make their lives easier and perhaps avoid some of the struggles they faced. I have summarized what they shared in a series of short case studies to inform the other chapters of the book as research does not always happen under the circumstances of reality.

Let us meet these adults who did not learn about their disorder until after graduating from high school. And please, as you read this chapter, learn from their successes and their struggles and share with the students and children you support.

Case Study 1 - Randy

What it was like to learn they had dyscalculia

Randy is a 68-year-old educator whose first real memory of something not being right was in high school. All her classes were honors except for math which caused a lot of distress and feelings of not 'being smart' and the need to hide it. This feeling of not being smart was pervasive and led to fears of being found out as not belonging to the right or the bright crowd. It was not until the age of 62 when she had to recertify her teaching license, and she took a class on math learning disabilities that she recognized that she was experiencing the same issues. Her belief was confirmed when she had a school psychologist conduct the needed diagnostics and confirm that she did indeed have dyscalculia. Her first reaction was one of relief, that there was an explanation for her severe struggles with math, that it was not because she was not bright, it was because her brain worked differently. But her relief gave way to anger as she dealt with the thoughts that she had avoided why she was different and that learning earlier could have made things easier.

The impact on their lives and careers

Randy described how she experiences dyscalculia as problems not just with performing math but also with colors. As she put it, "I can read and do other things at the same time. But if I need to do math, then I can only do math, the TV must be muted, and I need to focus solely on the problem at hand." And she added later, "The other thing is that it is the only area of my life that I triple check everything that I do. Everything else comes very naturally. But with math. No matter how simple the problem is, 9 plus 4. I do it twice to make sure I get the same answer. So, I need time, I need extra time." Randy also shared that finances are probably the most difficult part of her life that involves math, so she checks and double checks all her bills before paying them. She relies on her husband and family to help her determine if one credit card is better than another, or how much it really costs when buying a car, interest rates and monthly payments for three years vs four years making no sense to her.

Dyscalculia has had an impact on her career path as she chose not to pursue her Ph.D. because of the statistics and advanced math classes that would be involved. Impacts on social life are still felt as she avoids any conversations with friends or on social occasions that may veer into the realm of mathematics or how it is used as she cannot participate and feels that others would not necessarily understand. Can you imagine your friends discussing the scores of your favorite teams and none of it making sense?

But probably the most difficult thing was the impact it had as a child on her self-esteem and the math anxiety it caused. Her story of the 3rd grade math teacher who humiliated her because she could not memorize the multiplication tables and the geometry teacher "Who told me that he was gifting me with a D, because he just could not think of having me for another year in geometry. And it wasn't that I was bad or that; he just said he was so disheartened as a teacher that he couldn't help me see my way through it."

Strategies for doing math in day-to-day life

So, what strategies has Randy adopted to live in a world where math is pervasive? One of the areas that I found surprising because so many think that those with dyscalculia have such difficulties is that they will give up trying to master math. That was not the case with Randy. She had never really tried to do mental math but now that she knows what dyscalculia entails, she tries to solve problems in her head before putting them on paper and she practices the math facts while going through the car wash. She is improving slowly but surely. She visualizes TouchPoints on numbers, and she is working to understand percentages so that she can choose the right credit card next time. She has not found calculators to be particularly useful as she cannot tell if she has put the wrong number in. So, with them she double and triple checks her calculations. She has actually found adding machines and excel spreadsheets to be easier to use.

When it comes to the most common use of math today, driving a car, Randy is in luck – the original GPS devices told one to turn left in 700 yards and directions such as left, right and turn in 700 yards resulted in a lot of U turns for her. The newer GPS systems show you which way to turn and tell you that it is at the next light. The other ubiquitous numbers in our lives are our social security numbers and our ATM pins. She has learned to sing and dance them as the music breaks it into smaller chunks and the other sensory areas allow the numbers to be more easily recalled.

Probably the most interesting adjustment that she has made in order to do math is with cooking. Just think of the math involved - following directions in a set order, temperatures, time, fractions, etc. Randy has learned to rewrite all recipes in the actual sequence in which the recipe unfolds. In other words, she mixes the ingredients with the directions and premeasures everything before starting.

Their advice for the rest of us on how we can help

At the end, Randy had a few words of advice for parents and educators as she knows that her experiences with math, her feelings of being stupid, and the imposter syndrome would have been quite different if she had been diagnosed in school, and dyscalculia was as well-known as dyslexia. Her requests to teachers and parents, "let me use manipulatives, do not humiliate me, and do not say you will 'work with me.' Please understand me and give me the time to think through this."

Case Study 2 - Carrie

What it was like to learn they had dyscalculia

Carrie is a 38-year-old who has a clear recollection of doing well in math until third grade when multiplication tables were introduced, and she could not memorize them. As she described it, she "did not get it, it felt like there was something wrong with me in math and it just made me feel broken." Reading on the other hand was a subject that she took to and was reading books before she entered kindergarten.

Math was a constant struggle through middle school and worsened in high school. In her senior year, her pre-calculus teacher tried to help by having Carrie come in during her off-period time and worked with her. Carrie was lucky as her teacher realized something was not right and told her, "I will pass you because I can see you putting in the effort, she said, but for some reason it seems like your brain just doesn't get it." That was when she first felt like there was something wrong with her brain when it came to math. She excelled at everything else. It was not until her senior year that a teacher identified that there had to be an issue. Did it make sense for math to be a struggle? Definitely not. She was attending classes for the gifted and was in the IB program until she withdrew so her classmates would not know she struggled with math.

It was not until the age of 32 and for completely different reasons that she went through the diagnostic process and a variety of disorders were identified, dyscalculia along with ADHD and Autism Spectrum Disorder.

As Carrie described it, when the clinical psychologist explained how she was neurodivergent, she felt empowered as she now knew that her brain processed information differently, it was not that she could not learn it just had to be different. "I felt like, okay, I was no longer, I wasn't just broken and weird and nobody liked me, and I couldn't just do things that other people could. I actually had, there was a clinical reason, a scientific reason that my brain just at its core processed information differently than other people."

The impact on their lives and careers

Carrie dreamed of being a geologist for most of her K-12 life but was told as she approached decision making time that becoming a geologist would require taking a lot of math classes. Her change to art education was driven by taking as few math classes as possible. There was a bright side to this career switch as Carrie believes that by becoming a teacher, she was exposed to how math is taught and was able to incorporate those lessons into her daily life so that math became less difficult for her as she was able to learn how to learn math. She persevered and today is a certified tax trainer who teaches business owners how to do taxes and is licensed to do taxes for those who need help.

Her struggles with math also impacted her personal and social relationships. The homework struggles and being told to try harder made for a strained relationship with her parents and impacted on her willingness to join social clubs where her difficulties with math might become known. "I already was the weird kid. I was really intelligent, but very odd. And I didn't want to be teased more, made fun of more, give my peers more opportunities to call out that I was struggling in really specific areas."

Strategies for doing math in day-to-day life

Carrie has learned a lot of strategies that she can use to do the math that is needed for day-to-day life. These include using what she learned from TouchMath when she was a child - to place dots on numbers when she had difficulty working on a problem or recalling a fact. In addition, she does not have analog clocks in her home, only digital. But she has not acquired strategies for learning everything. She does not have a strategy that would enable her to cook and prepare meals has not been something she has been able to do. She relies on her husband to do all meal preparation as recipes and temperatures do not make sense to her. We joked that if he were not around, she would be a major takeout and order-in customer with the local restaurants.

When it comes to the straight math, such as calculations and facts, Carrie has improved but as she described it, she practices constantly but has no automaticity with certain fact tables, the 7's, 8's and 9's in particular. So, a calculator is necessary for Carrie, but she needs to work on the problem multiple times as she does not know when she enters the number incorrectly. The multiple attempts decrease the possibility of entering the numbers incorrectly. When she must work math in her head she will usually add or subtract from the 5's facts. So 7x6 becomes 7x5=35 and count another 7 to get 42.

Part of everyone's day-to-day life that requires math and is often overlooked is the math needed to drive a car. Judging speed, turning left and right, reading a speedometer, all are considered part of the math universe. For Carrie, none of these were easy to do so the advent and easy access to adaptive cruise control and GPS that tells what the speed limit is and uses colors to let you know you are exceeding the limit are lifesavers literally.

Being on time for meetings and daily life activities is a major struggle. Setting alarms has helped but reliance on calendar notifications has allowed Carrie to be ready for meetings. But she had not yet learned to use tools such as Siri and Alexa to set multiple countdown alarms in order to be on time for other things. For her, meeting friends for lunch is difficult as time has little meaning. Fortunately, her friends know why she is often late and that understanding allows the friendship to continue.

Their advice for the rest of us on how we can help

Carrie had some good advice for all of us. First and foremost, "look at the things we do well and start early with the child and start as soon as possible with parents and teachers." Her advice to them is to please be positive when talking to your child. If you think something is not right, remember that there is a logical reason for why a child can excel in one area and be totally lost in another subject area, trust your suspicions, and find out why. "Like, you know, they don't seem like they're five wise, 10 wise, but they can write an essay better than most of the rest of my third-grade class and they're avid readers."

She feels she lucked out with several of her teachers and is very grateful to them as they recognized her differences and reassured her that she was "going to be ok." But she wishes that they knew then what is becoming more common knowledge now, as she thinks she would be much further ahead in her career and not have had so many self-doubts. As a current teacher, she is practicing what she preaches and alerts parents when something does not appear right. The child who "is so smart and they do so well in other subjects, but math seems to be an area where they struggle the most. And, and it is strange because I noticed things about these behaviors, you know, they have a really hard time doing math facts. They have some struggles with even basic addiction or addition facts."

She is also a huge advocate of the MTSS system and its ability to help find and identify students who need special support. That is the teacher's expertise that can save a child today from some of the struggles and self-doubts that Carrie had.

Case Study 3- Elizabeth

What it was like to learn they had dyscalculia

Elizabeth realized that she did math differently early in elementary school when she would sit in class and daydream because, "*I just didn't know what they were talking about.*" When teachers told her mother that Elizabeth was just not very smart, her mother fought to get her help and by 3rd grade she had been diagnosed with having a learning, written expression, and auditory processing disability. The test was a memorable experience for her as for the first time there were no instructions, she was simply given a problem and had to figure it out, which for her was a great deal of fun. Because of her age she does not remember ever really being told, only that she had to start going to another class with many fewer students and most had behavior problems.

Elizabeth is one of the fortunate ones. Her parents were able to find and afford private tutoring and were also able to make sure that she practiced her math facts via games every day. So, for Elizabeth, her dyscalculia was discovered early enough that it was her normal and she had parents who were supportive of her need to spend more time practicing the foundations of math and receiving extra instructional support from a private tutor. To this day she still struggles and always will with not being able to memorize numbers and number sequences.

If she runs 28 miles she cannot remember the number of miles, it has to be written down and she constantly works to remember numbers of any kind, sequences of numbers such as her phone number and PINs.

The impact on their lives and careers

Elizabeth described how her dyscalculia has had a definite impact on her career choices as she knew that without math and passing certain classes, she might not have the career she dreamed of as she would not be able to go to college, much less graduate. Her dream was to be an archeologist. She was accepted to college but was unable to graduate as she could not pass the college algebra test. She attempted the class 5 times and had support from her teachers, but to no avail as at the time college level algebra was "*memorization of how you have to do each thing right- if you do one thing wrong, it's going to mess the whole thing up. You're not allowed to use calculators, so I was never able to pass that class.*" Elizabeth did not know that she could have informed the college that she had an IEP and probably received the support needed to pass the algebra class.

Strategies for doing math in day-to-day life

In day-to-day life Elizabeth has acquired a number of strategies that make it possible for her to successfully do her high-stress paramedic's job. Memorizing the address coming over the communications system when being sent to an emergency is not something she can do so she enters it into her phone, handwrites it on paper, and then double checks the accuracy as lives are often at stake. In her job, she has learned to advocate for herself and has no problem telling her colleagues she needs them to be quiet so she can take a blood pressure reading and remember the numbers as sound interferes with her ability to hear and remember the numbers.

She has also learned to use the 5's and 10's skip counting sequences used for money in order to determine drug dosages. Rather than seeing the math problem as one of multiplying by fractions tied to a patient's weight in order to determine the correct ratios, she equates it with the monetary system with which she has had more practice. The amount of medication needed is tied to how much money would be for each weight designation. This has allowed her to perform critical calculations in high stress systems. The use of the 5's and 10's has allowed her to develop estimation skills and has been particularly helpful for day-to-day financial processes. Rounding values up to the nearest 10 allows her to make sure she has enough to cover a bill, but it does not help in making sure that she has entered the correct number, for that she relies on the bank to check as what she writes does not always match.

Her technique for dealing with math anxiety is one of avoidance. Math is such a stressor for some daily activities, that it is easier to rely on her husband than put up with the stress. She also practices positive self-talk. "*Some of the self-talk that I give myself is I just always remember all the things that I'm really good at. And the thing that I'm not good at is really kind of small. And it doesn't affect me.*" She credits some of that to an old PBS film called "I'm Not Stupid" which helps her remind herself that she is not.

She also found other ways to deal with the anxiety and her low self-esteem which she believed were tied to her poor math skills. She believed if she could excel at something she would be able to feel better about herself. She decided to start track and field and was able to change from hanging out with underachievers as she did in high school and instead learned that she fit in with those who overachieve and are successful.

She has learned to take the time needed to ensure that information is stored in her long-term memory. To do this, she has to read a book, highlight the book, take notes, make flashcards, and then practice repeatedly. A highlight that she shared was that learning second languages has not been difficult for her and she wonders if that was because she does have issues with learning math.

Other adaptations and strategies that Elizabeth has used include extensive use of the calculator to double and triple check her work as without that she makes mistakes and is unaware of them. For remembering the numbers that are critical to our lives; PIN numbers, spousal and children's birthdays, etc.; she knows that these take time and effort to recall. Elizabeth also shared that being late is something that triggers a lot of anxiety so she sets all her appointments 15 minutes earlier so she can make sure not to miss an appointment. Issues with using cash are solved by pre-counting everything into consistent amounts such as all packets will be 10 dollars and then she can count by 10's. But counting $10.50, $10.75, $11.00, etc. is not possible.

Their advice for the rest of us on how we can help

The most important advice Elizabeth wanted to share with me was the advice she would have given herself if she could go back in time. "*I know I would just tell myself that you're still going to be able to be successful in life. You're just going to have to work a lot harder.*"

Her plea as well as advice to parents was for patience and understanding, that the challenges of doing math were huge. "*I would have liked to be rewarded more or even some when I did good because it was so mentally taxing. It was like a brain workout for me. 20 minutes is very overwhelming for a kid when they are challenged to the point that I was. I would have told them that ..., expecting me to practice for an hour a day just made my brain fry. It wasn't helping me.*" For her teachers, it would be to give more time to do the work and be more circumspect when handing out test papers as hers were so bad she was always embarrassed. When explaining the problems and how to work them, to please break them down into steps and share multiple problems with those steps as that helped her a lot.

When it came to her advice for the current generation of parents and educators, she commented on the assumptions she came across when sharing her learning disability. Because it appears to be invisible with few outward signs, she has noticed that there is a general expectation that a disability is obvious and recognizable by its behavior and cognitive manifestations. She has found that she must educate people about disabilities, especially dyscalculia, and has used her disability as an advocacy platform. Her closing advice was so practical that I am sharing it in its entirety "*I would tell them to find out what tools work for them.*

Like, if they have a calculator, use it. It doesn't matter if anybody else uses it, use it. Write things down if that's what works for you - a pen and paper. Actually, writing it down if that works for you. Watch videos if you have problems with cooking, get a cookbook that has pictures of every step. If you're having problems with speed or distance, get something that talks to you and tells you what you need to do instead of looking at it. Use every single tool you can with no shame because that's why we have them. We're not in school being tested on things. We're just trying to survive."

And advocacy, both for others and herself, was seen as especially important. "*And I think that if people don't know that you're having trouble, they can't help you. I think it's really important to tell people what you need, and that's something that I had to learn how to do...*"

Case Study 4 - Samantha

What it was like to learn they had dyscalculia

Samantha was not diagnosed until her 3rd year of college, at the age of 21 or 22 as she recalls. But her memories of struggles go back to kindergarten, where she realized that her classmates could say their seven-digit phone numbers, but for her, it was an impossible task. The introduction of addition and memorization of the addition facts triggered such severe anxiety that she had nausea. She shared that her K-12 math educational experiences were such that she managed to do just barely good enough that she was accepted to college.

But that is where she began to fail "just miserably. Everything was a struggle. I was depressed, I just couldn't keep up with everything despite going to classes and doing all of my work." Fortunately, she had a professor/advisor who checked with her other professors and realized that something was not right and put her in contact with the Student Support Center where she was tested. Her grades were so poor that the Dean gave her one semester to prove that she belonged and could succeed in a university setting. The Student Center expedited her testing, determined she had dyscalculia and provided services to enable her to access therapy, tutoring, study groups, etc., so she could successfully complete college. Samantha believes she was one of the fortunate few and the diagnosis was a relief as it explained why she struggled and still struggles to this day. She knows that if others had not identified that something wasn't right and sent her to the Center, she never would have thought to do so on her own, that dyslexia was so well known but not her form of a learning disability, dyscalculia.

The impact on their lives and careers

Samantha began her adjustments to her undiagnosed dyscalculia from the start of elementary school by recognizing that if she had a quiet place to work, such as the school hallway, and she could focus on one and only one problem, use her fingers and take her time she could answer questions correctly. She could not complete a full assignment but what she completed was correct. She also learned at the approximate age of 7 to blurt out any answer that came to her and she could get the teacher to send her into the hallway to do her work. Smart kid, I say!

She also believes that her career path would have led her to where she is now in marketing and communications but that her dyscalculia probably impacted the route she had to take. She stated, "So my journey may have been different, actually, it would have been different, but I think the destination ultimately would have been the same." Initially she had not considered college because "my guidance counselor told me that I shouldn't waste my parents' money going to college because I wouldn't succeed, that I should find some alternative career path, but then didn't suggest any. It was just kind of like you shouldn't do this figure it out, which is one of the things that led me to delaying applying for college." When she began attending college, it was a sorority sister that introduced her to the field of marketing and communications. Upon exploration she realized that it was where she was meant to be. Her comment that "if I had gone somewhere else, I may have been pushed into something a little bit different, something that I may not have liked as much as this. And you know, may have taken longer to find this, but I think I still would have ended up here." That statement is a strong indicator that Samantha has not let her disorder interfere with what she believes is her goal.

Strategies for doing math in day-to-day life

Samantha has learned a lot of workarounds for her issues with the math world. Time of day and dates were and still are an issue. The date 11/19 can be seen as 11/9 or 1911. Reading an analog clock or a calendar is difficult to do and even with digital clocks she can misread the numbers. The use of digital clocks helps but she has worked to accept that if her understanding does not feel right, then she has probably made a mistake. Her sense of time passing causes her to be late or completely forget an event as well as misjudge how long it will take to get a project done. Immediately committing meetings to calendars through writing helps as does setting a simple timer and working until the timer goes off. Another strategy she uses is time zone conversion apps as her meeting schedule must accommodate others who live in different zones.

Directions and navigation are another major area of her life where she has had to learn strategies that will enable her to drive and get where she needs to but getting lost is a definite issue for her. Distance, whether a block or a mile has no meaning, and turning north, south, east, or west also means nothing. So, for her, the GPS navigation systems on the phone and in cars have been a technology solution that has made getting around possible. As for the speed, she uses cruise control and matches the speed of the surrounding cars.

Cooking is a challenge, and she has learned to make sure that she concentrates and takes her time. If she does that her recipes work, but if she multitasks or does not focus, the results are a "toss up."

She has found acceptance of her need to still use her fingers as a means of calculating math problems in day-to-day life. Samantha commented that she is no longer embarrassed by her need to use her fingers to solve problems and that it takes longer than most of her colleagues to complete a problem. If others do comment on her difficulties, she explains that it will take a little longer, but she can solve the problem.

Dealing with the math that we all do for personal finances, credit card bills, car and mortgage payments is accomplished through setting up automatic payments and making herself do the finances and double and triple check them. She uses her Excel templates and keeps her accounts separate from her husband's so that when there is a mistake it does not impact them both.

She has also learned that stress makes completion of math problems difficult, and she must manage that stress. Double and triple checking her work before sharing has helped as has accepting that she and everyone will make mistakes. But these strategies decrease her anxiety. Tools, such as digital clocks and cruise control have made math easier for her and also help lower that stress. Her Excel spreadsheets where she can set up a template with the needed conversions and not need to worry about mistakes were included in that list.

Their advice for the rest of us on how we can help

Samantha had some good advice for us and how we should move forward to improve the recognition of dyscalculia and prevent more children having the issues that she experienced. In her own words, "I think the biggest misunderstanding is just that it is different from dyslexia, that it is stand alone, that it has its own category. So, people are quite surprised. But I feel like the people that I tell have so little knowledge about it that they don't have any conceptions at all and it's a blank slate. So, I get to educate them, which is pretty cool actually... It means if we get on top of this early to do, you know, really good education of the general public, we got a chance here."

If she had the opportunity to go back in time and talk to her 14-year-old self she would be encouraging and say "it sucks now; things are hard. You don't like school, you don't think it's fun, but there are people who believe you and believe in you and keep your head up and keep giving it your all because eventually you will love learning and you're going to find something that you absolutely love to do every day and you're going to have the opportunity to help other people and make a difference and it's all going to pay off in the end. Just keep going."

As for her teachers, she was sympathetic to the incredibly hard job they have but she does feel that there were those who had the opportunity to change her experiences of school and the memories are still there more than 20 plus years later; "... don't just automatically label me as stupid or lazy or tell me I shouldn't go to college. "

Her advice for parents was blunt, "You know your child better than anyone else. So, if you think something is off, there's a good chance you're right and advocate for your child. Don't stop. Keep asking questions, keep pushing for answers. Do your homework. If the school won't help, find people who will. They're out there. Sometimes you have to look for them, but don't, don't stop."

Case Study 5 - Theresa

What it was like to learn they had dyscalculia

Theresa's first real memories of learning that she had dyscalculia were in middle school when she was told that she had to go down the hall to the special teachers for her math classes. She believes she knew earlier as she has memories of not being able to dial a 7-digit phone number. She also remembers quite clearly the feelings of stupidity, frustration, and pain because her parents did not understand that she was struggling and did not try to seek help for her at a young age.

Her career aspirations were initially limited as her belief in her ability to learn math was nearly nonexistent. Working in a daycare center, where a college degree was not needed seemed her best option. As she put it, "Children don't care if you don't know upper-level math, they don't care. As long as you can read simple repeated line books and sing songs. My assumption was I would be a pre-K teacher or a pre-K paraprofessional which takes a 2-year degree. That's what I was in school for. Didn't take any math to do it." Even that aspiration was due to a high school teacher who recognized her abilities and literally made her complete the coursework.

But in Theresa's case her new husband changed all that when he told her that she was to get a bachelor's degree of some kind, any kind. A chance encounter at a university pregame party with an occupational therapist (OT) and learning it only took a basic college algebra class, she was sold. Her dream was to be a neurobiologist, but the math coursework required made it impossible but to become an OT and being able to work with the special educational population made up for it.

The impact on their lives and careers

Theresa has learned to respect herself and her capabilities, having the attitude that folks will "take me as I am or not." She described this attitude as a self-defense mechanism that is necessary in order for her to function. She explained, "I spent the first 17 years of my life trying to figure out what was wrong, why, why I felt so stupid, why I always felt like I wasn't good enough, and I never was." Her choice of initially working in a daycare was a result of those years of feeling "stupid" as the little ones don't care.

She has also adopted the rule of three in order to deal with the stress and anxiety that goes with her dyscalculia. "I'm allowed to do three stupid things every day and I have to forgive myself. All three of those times. Now if I get to the fourth thing, it's just time to pack it up and go home because something's just not right. But I allow it." Theresa shared that these feelings and her dislike of people not understanding them has led her to shun groupwork whenever possible, another coping mechanism that allows her to avoid labeling herself as stupid and decreases the anxiety she feels when she is not successful. Her work on her self-confidence is deliberate and sustained as she knows that math will be a struggle, and that people will not understand her difficulties.

One of the areas of life that Theresa avoids is playing games with friends and family unless it is a game that does not have scoring, or it is a solitary game. Her attitude is one of "I have no use for it because. I'm not going to win anyway, so why?"

Finances are done very carefully with double checking and triple checking. She sets an automatic withdrawal and then she goes to the statement immediately and verifies the amount is correct and then uses the calculator to do a final check.

Strategies for doing math in day-to-day life

For Theresa, mental math is not doable, and she accepts that and relies on the calculator. She knows the 2's, 5's and 10's facts and has consigned the rest of the math facts to the 'I don't care bin'. Her inability to hold the facts in her head or do any form of mental math has forced her to write everything down but even then "dots and commas meant nothing" to her when looking at them. This impacted her ability to maintain her department's budget, but the bookkeeper was able to allow her a $50 margin of error with the budget and show her how to aim for a zero-dollar balance. That is what the calculator is for even if she has difficulty recognizing if she performed her calculation incorrectly. However, she has adopted a strategy of looking at and saying the number aloud as she can then tell that there is a discrepancy as what she is saying does not sound like what she is seeing. This has allowed her to take the problem to another person and seek assistance, necessitating building a close-knit circle of colleagues and friends who help her with math.

She has been able to adapt to driving a car and her difficulties with telling directions by using her wedding band on her left finger as the differentiator between left and right when needing to make a turn.

Technology has been a large support for her as things such as speed dialing were life savers. She needed others to enter the number and store it for her the first time but after that she was good. Other supports she uses are various technology tools on her phone for securely storing important numerical information such as PINs, phone numbers, SSNs, etc.

For Theresa one of the most difficult areas of everyday math was cooking because the recipes and the mix of language and numbers: measurements, time and temperature were confusing. The results frequently being that an ingredient would be skipped, or the temperature would be incorrect, leading to less than edible results.

Their advice for the rest of us on how we can help

Theresa had advice for parents with children who have dyscalculia and that was about their child and that "they're having a hard time. They say I did something, or I didn't do something. The best gift you can give your child is to teach them how to forgive themselves." Theresa's parents were of the mindset that she did not try hard enough, and that more effort would overcome all issues. This extended from schoolwork through all of life's activities including helping to prepare the family meals.

Her main request for all of us was to expand awareness of this disorder and help those who have dyscalculia be comfortable with their different way of doing math.

Case Study 6 - Parents M & B

The last case study is with the parents of a young college student who suspected their son had more than just dyslexia and dysgraphia but did not get a diagnosis until their son began high school. I felt it was important to explore how parents' experiences determine if and when they seek a diagnosis, the support they get from the educational system and their advice to others.

What it was like to learn their child had dyscalculia

In the case of M&B because their names are distinctive and known in the educational world, I will only use their initials. M&B knew that math would be difficult for their son because of his ADHD, dyslexia and dysgraphia but thought the struggles would be associated with those disorders and had no idea until high school it was also dyscalculia. Their son's math issues began in late second and early third grade when it was time to learn the multiplication tables and he started lining his numbers up incorrectly and had other errors. In 4th or 5th grade the elementary school moved to a new math program, and the issues were made worse, but it was possible to attribute the struggles to learning a new method of solving the problems, one that was more conceptual and less a single procedure. The need for exploring a math problem and discovering a solution rather than the teacher explicitly sharing how to solve the problem and then doing it with the student's methods was a definite issue. At the time the struggles with math were attributed to the change in curriculum and the other disabilities.

Not until he was in high school, around 14 years of age and the routine three times a year IEP review was due that a diagnosis of dyscalculia was returned. M&B were determined that since he was entering the public system that he should have every accommodation he needed. They credit his counselor, who, even with a case load of 500 or more students, would be in contact with their son and communicate with them to ensure they were always in the loop. They also became very involved in teaching him how to advocate for himself so he could learn to be self-sufficient. For him, the move to public school, the additional diagnosis and accommodations, and the support of the teachers, counselor, and his parents made for success. He is currently attending college and succeeding, he has not determined his career path, but there is no reason he cannot pursue what he wants.

The impact on their lives and careers

M&B are educators, with one of them also being a special educator. So, from the start their son had an advantage. M&B knew how to seek out the accommodations and supports that would help him. This meant finding ways to engage him in day-to-day life. Sports were not a love but technology and being able to be social via that medium was an interest, especially with his nonverbal learning disability. It also meant teaching him how to advocate for himself with teachers and others in order to get the support written into his IEP. This included more time to complete work and asking his teacher for that time and suggesting where he might be when completing it, such as during his advisory time. It also meant that he needed to accept he had that extra time; he did not need to rush through the work and could and should ask for a tutor when he needed to prepare or was struggling. All of this has now carried through to college and hopefully on through his adult life, recognizing when his struggles are due to his disability and immediately seeking the support that enables him to work through it.

His parents also provided extra activities to help work on executive functioning, homework, and study skills. As he progressed through the grades the active support became less as the skills, he was learning went beyond the skill sets of normal parents. The parents who can teach high school math classes are rare so teaching him how to advocate for himself became the primary focus.

Strategies for doing math in day-to-day life

M&B have also been active in helping their son determine the day-to-day things he needs to do to use math successfully. This includes writing each and every step needed to solve a problem, verbalizing it aloud, and doing his work in a separate room so he could talk aloud. They also encouraged him to use a calculator, turn his paper sideways or use graph paper so he had built in columns for problems, and using fact tables to check his work. They shared that it was frustrating at times that teachers did not know these simple ways to support a child who did not yet understand or struggled to understand the concepts and procedures and they could not understand that all the students could not be on the same page of the curriculum, literally and figuratively for many reasons. They did find that flexibility to work math in a variety of ways increased as their son progressed through the grades. Some are due in part to the fact that math became more abstract and there were various ways to explain math. This is the difference between illustrating a fraction as just numbers or showing how it could be represented in real life through pizzas and other aspects of the real world.

Their son also had great difficulty with managing time, from telling what time it was to determining how much time was needed in order to prepare and arrive at a scheduled event. He learned to backwards map, building in plenty of time for preparation. The example they provided was "if he has to be at work at 5 o'clock he gets a shower at 12 o'clock, so he is prepared for 5 o'clock. He is very much a backwards planner, which is great. I mean, we all should be but that is very much a part of his of his day, kind of thinking about what's going on and then how he's going to work backwards to make sure he's got what he needs to be ready for that." The use of timers and alarms on his phone and watch are tools that have helped manage time and as is usually the case, digital is much easier for him to use than analog. It was also important to state the time, 5 o'clock and not 'you need to be there in an hour.' Time needs to be specific, phrases such as we will leave in a little bit, soon, awhile; have no meaning. Instead, time has to be made as concrete as possible, something that can be seen on the clock, watch, or phone.

Directions are still a struggle and at this time he is not confident enough to begin learning to drive. Anxiety about getting lost has made him delay getting his license. In this case, M&B are stepping back and letting him choose when and if he will learn to drive.

Their advice for the rest of us on how we can help

There were three areas of advice that M&B continued to come back to during our conversations. One was that joining a support group is important and provides a layer of reality. You are not alone and the parent or educator sitting next to you is going through the same thing with the same frustrations. You also can learn tips as to how to help your child. The second was that parents have an incredibly important role to play in the active teaching of the strategies the child will need to navigate the world and master the math needed to be successful. This can range from using alarms and timers to make sure they are on time or how to request the accommodations needed to successfully complete tasks that require math. The third and probably most important is to have faith in one's child and advocate for resources so they can learn and take their place in society.

So, what can we learn from the adults who have adjusted to life with dyscalculia and the parents who have walked alongside them?

There were some definite themes that ran through the interviews from which we can learn.

- The need to be shown explicitly how to solve a problem, to be able to work it in multiple ways and have the time to think it through.

- Career choices were impacted by the inability to pass college level algebra or other math classes.

- The common occurrence of significant others, namely teachers and parents, humiliating these individuals and to such a degree that the specific occasion is recalled with great clarity. The opposite is also true, the parent, teacher, or counselor who saw something in the child and encouraged them to forge ahead. The feelings of stupidity and/or being an imposter were noted throughout the interviews.

- Math anxiety was present in all of the interviewees, usually to a significant degree but all were aware of it and had strategies for dealing with the anxiety.

- Not knowing how to advocate for themselves at college and share that they had an IEP for other disabilities as that might have meant a dyscalculia diagnosis earlier, much less the high school guidance counselors.

- The use of calculators is not necessarily an answer due to the inability to recognize when the wrong number is entered. The 'dots and commas" issue.

- Many of the individuals developed well organized planning tools ranging from Excel spreadsheets that could do the major daily calculations needed for their professional lives to the use of times and alarms to plan for being on time.

- And last, but not least, being positive about their abilities to succeed.

There was so much that I took away from the time and experience these incredible people were willing to share with me. Each of them shared examples of the incredible challenges and difficult decisions that they needed to make in order to get to where they are today, successful adults who have adapted to their disability. What we take from their lessons and put in place today will make the current generation and future generations of children with dyscalculia have an easier time accepting and adapting to their disability. Just as no one today would dream of telling someone who is near-sighted or has dyslexia that they cannot be a doctor or engineer, the same should hold for those with dyscalculia. Let's look deeper into the disorder and what we know about it and how to provide the supports and education that can make that possible.

PART IV

In this part you will learn:

- Why it takes a team to support dyscalculia
- Normal milestones of math development
- Where students with dyscalculia struggle
- Math anxiety and how to address it
- Developing and implementing a support plan for the child at home and school
- Typical supports and accommodations that can be found in IEPs and normal life

Chapter Nine

SUPPORTING STUDENTS WITH DYSCALCULIA

So, what have we learned from the adults who were willing to share their experiences growing up with dyscalculia? Hopefully, you have noted some common themes and phrases. Those early experiences have a major impact both positively and negatively. Positive experiences include the knowledge that what happens in early childhood, such as early exposure to math – vocabulary, numbers, counting, having the math in the world pointed out, knowing that they are learning and doing math - have a lasting impact on academic achievement, high school completion, career choices, and mental health to name a few.

Without those positive experiences we have adults who describe feelings of stupidity, of not fitting in, of being an imposter, incredulity that they could read well but numbers simply made no sense. The student who struggles with arithmetic is a conundrum to not only teachers and parents, but usually to themselves. The student can usually read at grade level or above, does well in other subjects and receives the same classroom instruction that their peers do. However, math, particularly remembering math facts and any form of calculation are a constant struggle, leaving the student frustrated and in their own words; "stupid." The student does not struggle with all math, usually only the areas that involve arithmetic, facts, and calculations, which can make all math more difficult and frustrating. This student may be a student with dyscalculia.

Because math is so important in our daily lives both personal and work, it is critical that we find ways to support the estimated 2.75 million students who have dyscalculia that is severe enough to need specialized educational support and the additional 5+ million who have some of the risk factors and will have some issues in regular classrooms. We see support for students who wear glasses or have dyslexia or need wheelchairs as something that levels the playing field so they can participate in life alongside anyone, but we are not there yet with this math related learning disability.

As parents and teachers, we imagine the careers our children will have, doctor, teacher, electrician, arborist. I cannot think of a job that does not require the use of math in the actual performance of the job. And let's include getting to work on time, navigating the streets and understanding directions. All mathematics. Now add in that it is projected that there will be, and 2 million of those jobs will go unfilled, despite a national focus on STEM over the past decade.

How many of these jobs are going unfilled because we have students who do not think they can do math or have not had the benefit of the supports that make it possible to acquire the skills to be engineers, doctors, electricians? As you could see from the adults who were interviewed, for some, dyscalculia is a bump in the road, but for untold numbers who were never diagnosed it is probably an insurmountable barrier. The adults I interviewed saw it as a barrier they could overcome, and they felt it was much easier to navigate with support from educators and parents rather than on one's own.

If you read Part 3 and the Case Studies, you will have noted that I talked with successful adults. These individuals almost universally saw themselves as lucky; they recognized their struggles and overcame them through sheer grit and sometimes with the help of an adult who saw their potential. I am currently looking for additional adults to interview and it is not easy as so few individuals have been diagnosed at this point in our educational history. Few are aware that they have a specific learning disability, much less pay the fees to be diagnosed once they are adults.

So, what can we do as adults; the educators and parents; in our children's lives; especially those who are struggling?

We can start with looking at our own views of math and our previous experiences with math and putting those away. We know so much more about how we learn and the role that our views play in our children's learning. This is especially important in understanding how to support a child who is struggling with math or has dyscalculia, understanding how children develop into math-able adults and how that impacts the acquisition and development of math.

How the adults can support the child with dyscalculia

If you are like the majority of adults, you may not feel capable of supporting a student struggling with math or with dyscalculia. This may stem from your own feelings that you are not a trained educator so you don't want to confuse your child by teaching them differently than is being done in the classroom, that you never really understood math yourself, or you may have your own case of math anxiety.

But there is so much you can do as a parent or adult to help your child. I've collected some simple things to do that will go a long way to be a positive part of the journey. The key thing to think about is that **something is better than nothing** so think positively about math and talk positively to your child about math. You can and should:

- Learn more about the normal milestones for a child and math development. I go into detail for you in the following pages.

- Learn about the 4 main math domains that students with dyscalculia normally struggle with and the more general areas such as executive functioning areas that make math possible. These areas are number sense, learning the math facts, doing the calculation, and mathematical reasoning. In addition, there are the more general areas of education that are impacted; working memory, attention span, processing speed, phonological processing, spatial skills and nonverbal reasoning.

- Learn about math anxiety and its impact on a child who has a math disability and so is guaranteed to need to work harder than their peers. You will need to examine your own feelings about math and the statements you make to your students and children about math.

- Develop a plan to support your child if you suspect they are struggling with math or may have dyscalculia.

- List questions to ask as you seek guidance and support for determining if your child has a math disability, what to ask in the meeting after the diagnosis and what supports and accommodations are appropriate for your child if they need an IEP or 504 plan.

- Play an active role in your child's math life. This isn't going to be a short journey; it will be part of your role as a parent for as long as you live. A good start is looking for the fun in math and all the easy ways you can do math at home and in your day-to-day life. I will share examples of what those are and we will go even deeper in Chapter 6. We know that what happens at home to support math predicts math achievement at school just like it does with reading (McCoy et al., 2018).

These simple things are a huge factor in children seeing math as a part of life and something they do all the time.

Chapter Ten

MATH MILESTONES AND WHERE STUDENTS WITH DYSCALCULIA STRUGGLE

The milestones- how does math develop?

Our children follow natural developmental pathways – they crawl, then walk, run, and then jump. As they age, the movements become more coordinated and quicker. There are natural developmental pathways for reading and for math as well. These learning pathways follow a natural progression, moving from simple to more complex over time, through each of the various domains of math and the skills and concepts in each domain that children are expected to master in PreK-12 schools. The domains are how we describe the large ideas of math, how we organize the math and build out state standards and when they should be taught during PreK through graduation from high school. They are generally accepted as being:

- Numbers and Operations
- Algebra
- Geometry
- Measurement
- Data Analysis and Probability

If you are an educator, you may be thinking of the major domains as:

- Counting and Cardinality
- Operations and Algebraic Thinking
- Numbers and Operations in Base Ten and Fractions
- Measurement and Data
- Geometry
- Ratios and Relationships
- Number Systems
- Expressions and Equations
- Functions
- Statistics and Probability

Math is something we are born with. Three-month-olds can recognize changes in the number of items they have been shown. Birds, insects, and other animals also have this ability which is actually related to survival. If we are looking for food, it is important to recognize that the tree on the right side has more ripe fruit than the tree on the left or in the case of a bee that one field has more flowers with nectar than another. Counting all the fruit to decide which tree has more is not an option. But humans with our more developed brains continue to add math skills enabling us to solve more complex problems, doing them more accurately, with more flexibility around ways to solve the problem and our speed of completion increases. In math the development of these skills follows a standard pathway that all children follow and are clustered around the main areas of math that progress over time but in parallel. These areas are numbers and operations which are the foundation for algebraic thinking. Geometry or recognizing shapes and that shapes can be made up of other shapes, measurement which is used throughout our lives, data analysis and probability are the other main areas.

All children follow the same pathway or learning trajectory to master these as they are the core of PreK-12 math education and are the basis for the math we need in daily life. But each child follows a unique learning trajectory, they move at different speeds and with different supports as they move through the different stages of mastering the areas of math, the 'math progressions'. We also know that math develops in a hierarchical fashion that builds steadily to more and more complex skills and procedural abilities. The illustration below is from the National Council of Teachers of Mathematics and shows their recommendations on the emphasis each of the big areas of math should receive in relationship to the others over a child's PreK -12 career. NCTM - Executive Summary Principles and Standards for School Mathematics (accessed 8.30.24).

Most children will follow a normal developmental pathway for mathematics unless something interferes with that development, and it happens before the child is born. Our brains are as unique as our fingerprints, even identical twins have unique physical configurations to their brains. There are unique regions of our brain that are responsible for "doing math" primarily in the parietal lobe but there are also bundles of white matter- the neural pathways - that connect the math regions to areas of the frontal lobe that tune out interference and let us concentrate, that keep anxiety at bay, that enable us to remember that left is on that side or that 12 stands for 12 of whatever object or item we were just told to count or to look at. This is what neuroscience has taught us, where the grey and white matter is different and is preventing or interfering with a child's ability to learn and perform math like most children.

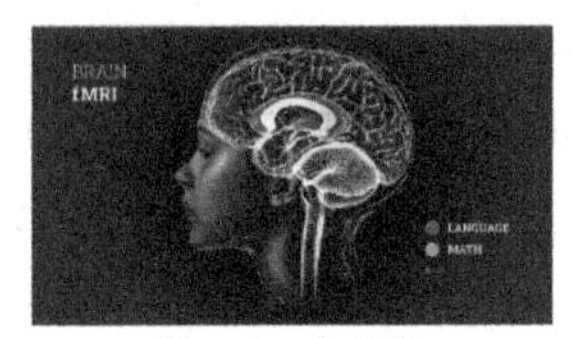

Brain fMRI

How we develop mathematical thinking and how it is different for the child with dyscalculia is important to understand. How does a 3-month-old 'knowing' that when you took away one of the objects behind their blanket, there was a change in quantity; and how does that same child become a 5-year-old adding 3+2 and getting 5 and then mature into an adult who is determining if they can afford the rent to a new apartment?

The window of opportunity for the most effective interventions is early in a child's life but it never closes, it only takes more time and is harder as one ages. But we do know from the research around neuroplasticity that with interventions and the right supports, the brain changes, it is plastic. The connections between the areas of the brain responsible for doing math and the speed and accuracy with which those math processes can happen change with deliberate training. This translates into comparable math performance for students with and without dyscalculia (Gilmore, 2018).

Are the changes in the neural pathway and the memories permanent? That question is yet to be definitively answered by science. But then, which one of us can parallel park perfectly if it has been 10 years since our last attempt to squeeze into a spot that should only be used by a motorcycle.

Those with dyscalculia struggle more with number sense, operations, facts and reasoning so support for those students should focus on those areas, especially in the early years. These four math domains are found early in the math progressions as part of numbers and operations and algebraic thinking. Due to the hierarchical nature of math, these domains are foundational to much of later math. My point being that math development for all humans follows a strict sequence and in education we teach that sequence. Feel free to grab your state standards and you can see the order in which these skills develop from K all the way through graduation. And remember what they say about those that start or fall behind? They almost always stay behind! That is what we are trying to avoid by starting early.

As with all things pertaining to how we develop as humans, the development of our math skills also has variability as to speed as to when one child learns to count 1,2,3 and another but there are general rules. The early years are critical for building the foundations for math as math is a hierarchical subject and it is essential that students master the early steps. If they don't, they are more likely to have compromised understanding in the future.

As you saw in Chapter 3, adults with dyscalculia have developed ways to adapt to their disability but your goal is to make it so there are fewer adaptations needed, or they are seen more positively as with my analogy of wearing glasses. No one thinks anything worse about those who need to wear glasses.

Let's talk about the four main areas of math that a child with dyscalculia will struggle with: number sense, the math facts, the four operations of addition, subtraction, multiplication and division, and mathematical reasoning and how they develop. I am not going to go into great detail as there are books written about how children develop these concepts and use them to build the higher order concepts of math. But I do want you to understand that these develop over time for all of us, but for the child with dyscalculia it will be slower as the regions of the brain that are impacted by the disability are the ones that perform these math areas. We will talk about interventions in Chapter 6.

Developing number sense

Number sense is basically our ability to understand numbers, their relationships and use them flexibly, not just memorized algorithms or the formulas to solve math problems. Number sense is the primary building block of all math. The operations and math facts are just what they say, quickly and accurately responding to 3+5 as 8 and 6 X 7 as 42. But number sense is broader and can be thought of as making sense of our number system and the four operations of addition, subtraction, multiplication, and division, and the ability to determine the most logical way to solve a problem or that the answer just does not look right.

We are born with number sense. Children as young as 3 months will react if they see 2 bunnies placed behind a blanket and one if is removed without their knowledge (Izard, et al., 2008). Their reaction is focusing their attention on the situation with one bunny as they are aware a change in quantity has happened. 2-year-olds can confidently hold up 2 fingers to represent how old they are when asked or become upset if their father gives them 1 cookie when they asked for 2.

This ability to instantaneously recognize how many objects are in a group and name it without counting the objects is called subitizing and it is thought to extend to 5 unless the objects are arranged in a pattern, such as found in an arrangement of 6 dots arranged in 3 rows of 2 or the patterns commonly associated with dice and playing cards. This ability to 'see' the 6 dots as a part-part-whole relationship and rapidly recognize it as 2 + 2 + 2 = 6 or 3+3 = 6 lays the foundation for memorizing the math facts, developing an understanding of how numbers relate to each other and understanding the operations.

Number sense is the ability to see and use math efficiently, in many cases doing it as mental math. Understanding that a number with a nine can be rounded up to 10 and then the multiplication is easier, and you subtract the ones. 9 x 15 can be worked as 10 x 15 for 150 and then simply subtract 15 to have 135. And to understand that if my answer was 200 or 1500 then I have obviously worked it incorrectly.

Poor number sense is seen when students choose to follow a procedure step by step instead of finding ways to make logical shortcuts based on their understanding of how numbers are related. Add to that, that those with dyscalculia have trouble recalling basic facts because they have not always made the connection between the numerals having a meaning. That 4 X 8 means 4 groups of 8 and seeing in their heads an image of 4 groups of 8 items. Instead, they are seeing symbols, A x B =

and there is no numerical meaning.

Examples of number sense are:

- Subitizing or recognizing the number of something without counting because it is a small number, or the pattern is a learned one. Can you tell me how many? Do not count the circles.

- That the numbers exist in a fixed order- 1 is always before 2 and 2 is always before 3, etc. One two three four five

- One-to-one correspondence or 1:1 correspondence, each item is only counted once

- Cardinality – that the last number counted is total number in the set or group

- Hierarchical inclusion-that the last number counted includes the numbers before it

- That number and quantity are related- Counting and understanding that the number words refer to a specific quantity and 3 dogs and 3 trees are always the same quantity regardless of whether the objects change. Having a mental image of 3 objects whether you see the symbol 3 or the word three or hear the word three.

- Conservation or that 5 objects can be grouped in any manner, but it is still 5 objects, position or change in a characteristic such as size does not change the quantity

- That numbers exist as symbols and words and have meaning- the symbol 3 and the word 'three' mean the same thing and that there is a definition attached to the word and symbol

- That numbers are counted and the existence or ordinality and cardinality. Understanding that numbers go in a certain order- 1,2,3 ... called cardinality and it does not vary; and that they represent placement – first, second, third.

- That magnitude and comparison of magnitudes exists-Understanding more, less, smaller, larger, greater than, less than, decrease, increase
- That a pizza can be small, medium, or large and the proportions are different
- Understanding that a number usually is made up of other numbers, the number bonds. 6 is made up of 6 and 0, 5 and 1, 3 and 3, 2 and 4.
- That numbers can form patterns, and it is possible to determine what number is missing- 3,6, 9, ?, 15 and the missing number is 12
- Being able to estimate, if the problem is 12+23 the answer should be closer to 50 than 100
- Understanding the place value system - that 10 is made up of 10 single units
- Being able to round up or down accurately
- Having alternative ways of working problems. If the problem is 24 X .25 it can be quicker to halve it twice as we are looking for half of a half rather than doing the multiplication problem.

There are many activities that can be done in the home or the classroom to help your child acquire number sense skills and these can be found in Chapter 6.

Developing the basic operations of addition, subtraction, multiplication, and division

For the child with dyscalculia those symbols that are the basis of addition, subtraction, multiplication, and division can be meaningless.

The +, -, x and / signs are not connected to the process that they or the words add, subtract, multiply, or divide indicated. 6 + 7 may mean that there are 6 objects and 7 objects, but I do not understand that I should add them to have 13 objects. An additional frustration can be that there are multiple signs for multiplication, X, *, the parentheses in multiple steps problems 3(4) + 7 = and the dot. Division can be indicated with a ÷ or the /.

The operations start with addition and subtraction as children recognize that there is a change in quantity. When I am given 2 toys and then another is added I have more. If two are taken away, then I have less. As I grow older and my knowledge develops, the ability to count, I think about adding 1 more toy to my current toys. I can count the first 2 and then I keep counting as I have 1 more and I now have 3 toys. I had to "count all" of them and as I expand my number sense mastery I can start with the number in my group and "count on" from there. I will not get into the mathematical laws that these constitute but your child's teachers and the curriculum writers follow the developmental progression as 'counting all' is done before 'counting on" This is especially reinforced by the child's development of the concept of part-part-whole. That the group of objects, let us say 5 is still 5 whether I am given 1 toy first and then 4 more or 2 toys and 3 toys. It would look like this:

- 1 toy and 4 more toys.
- I would count all of the toys to get 5.
- If I had 2 toys and was given 3 more toys, I would still count all of them in the early stage or my development.
- I would count 1, 2 and I would look at the new toys and count all of them. 1,2,3,4,5 because I do not yet have the number sense concept of the new group did not change the number of the original group so I could start counting from there.

It would not be unusual for the child to use their fingers and should not be discouraged. It is a normal part of the progression in the development of number sense. Over time the mastery of the math facts become a more efficient and faster means of solving problems but the fingers, whether it be a full-fledged count my fingers or the subtle finger touching many adults do to recall a math fact all is part of the normal progression.

Subtraction is the reverse of the addition operation and from there children move into multidigit addition and subtraction and soon repeated addition which is the beginning of multiplication and then onto division. For the student with dyscalculia this understanding does not appear to hold, they do not see subtraction as the inverse of addition or division as the inverse of multiplication.

Developing the math facts

Students with dyscalculia also struggle with math facts which are the single digit combinations for all four operations - addition, subtraction, multiplication, and division. Since we use 100's chart in most US curricula, for sake of our conversation let us assume math facts are 1-10 as everyone knows that zero is magic and is always zero. That means between addition, subtraction, multiplication, and division there are 400 facts that we want children to be able to answer within fractions of a second of being asked. The advantage to being able to answer instantly or be 'fluent" is that I do not have to work that part of the math problem so less working memory is used.

The child with dyscalculia will need longer to memorize the math facts and from what the interviews shared in Chapter 3 seem to indicate, most will never have all 400 but they all keep working on them.

I have some interventions for you to help your child work on the facts and the word "work" is deliberately chosen as this will be something that appears to be a lifelong project. Just like being able to really do parallel parking in 3 easy moves. Unless practiced regularly, most of us need 4 or more moves to parallel park and we usually think out loud or ask the others in the car to be quiet for a few moments as we remember and execute the steps.

Mathematical reasoning and how it develops

What is mathematical reasoning and how can I help my student develop it? It is a skill that makes it possible for an individual to use all the other skills they have in math by using logic to determine how to solve a problem. It is making sense of what a problem is about and the easiest way to solve it and if there are other ways to solve it, and if the answer makes sense. When a child explains their reasoning as to how they solved a problem, you can tell if they understand the problem, that they have thought through various ways to solve it rather than just following a formula or memorizing a math fact. An example is the fact so many adults seem to struggle with, 7 x 8. If you have not memorized or have difficulty recalling that fact, instead of panicking, you recall that 7 x 7 = 49 and count on from there, 49, 50, 51... Oftentimes the fact will resurface but you had an alternative way to get the answer that was logical.

Vocabulary, vocabulary, and more vocabulary - math can be understood as having 2 main components- understanding the concept and then using it in a procedural way to solve a problem. But underlying this is the need to understand and that means vocabulary. Even with the estimated comorbidity of dyscalculia and dyslexia it is still likely that your child won't have language problems and being able to acquire the language of math is a potential strength which will allow them to explain their reasoning and make it easier for you to determine if they understand the concept or know how to use the procedure.

You can help your child develop reasoning skills by asking why. Present problems and have them figure out how to solve them. Tell me how you got that answer, are there other ways to get that answer, can you show me how you did it? They will use this same type of thinking and reasoning in their science classes. Many parents may be hesitant to ask their child how they got the answer, but this is an area that can be a great modeling activity, and parents and children explore the solutions together. If you are hesitant, your child's teacher will be happy to give you some simple ways to do this.

Using your computers and phones to search for how to solve a problem and watching the many little videos that are common on sites such as YouTube make for a wonderful way to model how to learn, how to find answers, and that there are usually multiple ways to solve even the simplest problems. As adults many of us have forgotten how we actually learned the earlier math concepts.

Chapter 6 has more ways to help your child develop their reasoning skills and a lot of it is fun.

Typical Milestones and When They Develop

What are the typical milestones we see in children who do not have dyscalculia or other disabilities that could impact math development? There are many resources you can consult but typically they are as follows.

3-Year-olds

- Is starting to count but uses 123579 as one long string and does not distinguish each as a separate numeral
- If asked for 3 cookies can grab 3 cookies without having to count them out
- If asked how many, can answer 3 cookies
- Can tell you which group has more than the other group
- Can tell you who is 1st and who is 2nd.

4-year-olds

- Can count to 10
- Can count 10 items in order but may miss one
- Starting to understand that the last item counted is the number of items they have-this is cardinality so can answer the question "how many are there?"
- Is starting to write their numbers and can read them

- Can recognize 5 items without having to count them

- Can recognize whether 2 groups have the same number of items

- Can begin to place numbers accurately on a number line – 5 will be further along the number line than 2- this is ordinality, the consistent order that numbers follow, 5 always follows 2

- Can count all when given 2 small sets of items- 3 blocks plus 4 blocks

- Can backward count to see how many are left

- Is beginning to count on- 4 blocks + 3 blocks is seen as starting with 4 and then counting 4,6,7. I have 7 blocks

- Can show with manipulatives or drawn representations that 3 blocks need to add 4 blocks if 7 blocks are needed

- Can equally share cookies by using the one for you one for me strategy and knows that the odd numbered item must be divided, if possible, to be shared

- Dot patterns such as dice, dominoes, cards are easily recognized

5-year-olds

- Can count more than 10 objects

- Can count more than 10 objects when not arranged in a pattern

- Can draw the objects

- Understands cardinality- that the last number counted is the number of objects

- Understands ordinality- 1st, 3rd, 9th, and 10th

- Can answer what number comes after 7 or before 6

- Is able to draw a number line and place the numbers in order
- Does not confuse the size of objects with their numbers so can now compare a group of 6 mixed sized circles with a group of 7 mixed size circles and accurately identify the group of 7 as having more circles
- Can do single digit addition and is showing knowledge and fluency of math facts. 3 + 4 is instantly answered as 7
- Consistently uses part-part-whole thinking to do addition and subtraction. 7 can be broken down into 6+1 or 3+4 and is the same as 7-1 and 7-4
- Can recognize fractions when asked what is ¼ or ½

6-year-olds

- Counts to 100 with few if any mistakes.
- Knows doubles; 4 + 4 = 8, 6 +6 = 12
- Skip counts by 2s, 5s and 10s
- Knows all of the ways or number bonds to get 10. 10 + 0, 1 + 9, 4 +6, etc. and knows they can be reversed. That 6 + 4 is the same as 4 + 6
- Sees addition and subtraction as inverse or opposite operations
- Is beginning to multiply and divide
- Uses repeated addition and skip counting to multiply
- Uses grouping to divide
- Is starting to know the number bonds to 20.
- Can solve some unknowns such as 5 +? = 8

- Understands place value and that a number in the 10's column is a multiple of 10
- Can do 2-digit addition with carrying
- Understands that fractions are written as denominator and a numerator and the bigger the denominator the smaller the fraction
- Time awareness develop as well as measurement

7-year-olds

- Can count forward or backward from 100 and by 10's or 5's
- Can confidently count past 100
- Uses place value knowledge to identify which number is larger or smaller by looking at the 10's digit.
- Can estimate or round up

The above are the key milestones for children 7 and below. And nearly all concepts and procedures are built upon these. For the child with dyscalculia these are many of the areas where they have the most struggle due to the neurodevelopmental differences in their brains.

Why is it so important to start as early as possible? Because dyscalculia is not curable, individuals with dyscalculia will always process math in a unique way but it does not mean they cannot do math only that they will need to learn to use supports to do so. They will move through the math progressions in the same order, but they will move at a slower speed and need more support to master the concepts, skills, and procedures than their typically developing peers. There is still much to be learned about dyscalculia and the research is increasing our understanding and more is being done each year. But we do know that early intervention enables those with dyscalculia to learn and build on the early stages of math so that they can successfully do math with their neurotypical peers.

Math Difficulties in Adulthood

As parents and teachers, we ask ourselves, what is the future like for my child? For school age children, dyscalculia impacts academic success, mental health, self-esteem, and anxiety levels. For an adult, dyscalculia can impact career choices which impact standards of living if career choices impact salary which impacts housing, safety, self-esteem, mental and physical health (Vigna, 2022). Here are a few examples of what the adults I interviewed said dyscalculia was like for them and what you are hoping to prevent or make less unbearable. What do adults commonly describe?

Driving- how do you trust yourself to drive if you need to gauge your speed by how fast everyone else is going because the numbers on the speed limit signs have no meaning. What if the direction from the GPS says take the 3rd exit in the traffic circle or turn left and directions and numbers have no meaning. What if the only way you know a right turn from a left is that your wedding ring is on your left hand, so you touch the ring to remember how to signal.

Time blindness- this is a common way of describing having difficulty perceiving or managing time. What if getting to the right place is difficult, but now it also has to be at the right time, but 60 minutes and 30 minutes feel the same. Time management is key to punctuality today but also scheduling critical tests for health. Time blindness can also include procrastination, getting so absorbed in an activity that you lose track of time as well as underestimating or overestimating how long it will take to accomplish a task.

Financial literacy- What do you need to determine if your budget will stretch to include new shoes as well as coat because everything is discounted 20% and you have $150 to spend. How do you remember that the mortgage is always due on the 5th, and you have to keep enough money in the account to pay it if commas and dots do not make sense.

What if making dinner means laying out all of the ingredients ahead of time and writing out a step-by-step recipe that combines the normal how-to directions with the amounts so that a stick of butter is used twice instead of once. Just what is a pinch of salt?

What if you hear your friends talk so easily about negotiating a car loan or finding a great discount and saving $100 when to you all you feel is "I don't know what they are talking about!" These conversations always bring back memories of being told to try harder, that if you tried harder, you could get it! Imagine the guilt and frustration those words trigger.

Fortunately our brains never stop learning but it usually will take longer to learn new things as there is inaccurate knowledge to replace and gaps in early learning, think of trying to learn a new language when you are 30 or 50. So it makes sense to start early but the window for learning does not close so if you are reading this and your child is 8, 12, 15 or it is you and you are an adult just know not to give up but like most things it will take more concentration and more time. But the latest research in neuroplasticity supports that focused, repeated, and varied activities consolidate learning, cause physical change in the neural pathways that process math which in turns enables better math performance (Afolabi, 2024; Zacharopoulos et al.,2021).

The Development of the Domain General Areas of Math

The development of the foundational skills of number sense, facts, calculation and reasoning involve not just these domain specific skills but also domain general areas such as executive functioning of working memory and visuospatial (Fuchs, 2005; de Smedt et al., 2009). The ability of our working memory to both hold for short periods of time the visual of a math problem, any auditory information, and perform the mathematical process is critical to being able to do math. The visual component includes holding in short term memory the texture, shape, location, or position of components of the problem whether they be the string of numbers, and the operations involved or that the shape is a trapezoid rather than a square and what that means for properly working the problem. Other areas include to pay attention to what the problem is asking for and block out extraneous actions from the environment, information in the problem that is not pertinent such as the color of the boats, and potential solution strategies that are ineffective. Go back to that game of Simon Says in Chapter 1 and think about what has to be ignored in order to win at Simon Says.

Imagine the following word problem being shared verbally by a teacher, “What is the area of a rectangle whose sides are 14 inches by 7 inches?” Verbal working memory holds the information long enough to transcribe the problem which may include a sketch of the rectangle and it’s two sides or the individual may attempt to solve the problem with mental math and hold the picture in memory long enough to multiply 10 X 7 for 70, hold that in memory, recall that they must multiply 4 X 7 for 28 and then recall the 70 and add it to the 28 for 98. Then respond that the rectangle had an area of 98 square inches. The stronger the working memory and its ability to hold the numbers and the shape and work the math the stronger the mastery of the concept and ability to move further in the math progressions.

Chapter Eleven

STRUGGLING WITH MATH - MATH ANXIETY

If you ask most adults what their least favorite subject was in school and the pop quizzes and tests that tied their stomachs in knots; they would most likely respond" 'MATH!" Or tell an adult that the IRS called, and they are wanting to meet to review how you got the figures for a tax filing from 2020. I guarantee that 99% of us would have a bout of math anxiety. Math anxiety is an incredibly important issue to address, especially for students who are struggling and know that they do better in other subjects but for some reason not math. And it is not a rare feeling but one that is widely reported across multiple countries.

The feelings of anxiety, panic, paralysis, inability to think and the desires to avoid anything related to doing "math" are known as mathematics anxiety. Usually they are not mild feelings, they are negative and extreme. In younger students it is often described as due to fear of failure, difficulty, lack of time to complete work, and fear of bad grades (Szczygieł, M., & Pieronkiewicz, B. (2021). If we think of our friends and family, it is highly probable that we know at least one person who becomes anxious at the thought of figuring out a mortgage, taking a math test, having to explain how to do a math problem. These feelings of math avoidance at all costs and anxiety can play a huge role in general well-being and career choices especially as the child becomes older. Note that it is not limited to those with dyscalculia. Indeed, it is possible that dyscalculia is mistaken for anxiety.

We also believe that math anxiety and math achievement are related as they form a self-feeding cycle with anxiety impairing math achievement and poor achievement leading to ever increasing levels of math anxiety and so on. (Barroso et al., 2021; Aldrup et al., 2020). This in turn can lead to avoidance of classes that focus on math in school, college, and vocational training.

Current research estimates that between 20-30% of students experience math anxiety or other academic anxiety that has an impact on their achievement (Buckley, 2020; Williams, 2024). In the early grades, a child's math anxiety may remain just a collection of symptoms that manifest when it is time to do math, but it can be severe and show itself as physical symptoms. It is visible to parents and teachers as:

- physical symptoms such as clammy hands, rapid breathing, sweating, nail biting, emotional upset.
- avoidance of math games,
- stalling instead of starting to do the math work,
- failure to do homework
- avoidance of starting the work,
- doing the easy problems or not finishing,
- being slow to do the work,
- answering way too many questions with I do not know,
- not being able to show the work using the build it, draw it, write it process- shows confusion and lack of understanding, or
- trips to the nurse or bathroom when it is time for math or a math test.

Also, please do not pull out a timed test as those are the most anxiety causing. The older one becomes the more engrained the anxiety becomes and the need for more specialized remediation. Persistence into adulthood is not unusual.

Research suggests that math anxiety may be transmissible from adults to students, and teachers with anxiety can have fewer math interactions with students due to feeling less comfortable with teaching math. (Park et al., 2024; Prado, 2019; Schaeffer et al., 2021; Zhang, 2023) But research is suggesting that if you are a math-anxious parent, your frequent positive interactions with your child and talking about your own anxiety can help to mitigate issues of math anxiety (Guzman, 2022).

What causes math anxiety?

There are a number of identified causes of math anxiety- everything from timed tests, performance anxiety, lacking a growth mindset or can-do attitude, fear of failure and inadequacy to actual disabilities in the ability to do math-namely dyscalculia. Math anxiety also has a sociocultural aspect that is exacerbated by commonly heard statements such as " I wasn't a math person when I was your age" or "I didn't like math either when I went to school so if you bring home a C- or D and give it your best, we'll be happy." And it is still common to hear "some people are born good at math" or "girls aren't as good at math" or "math is not as important as reading." Parents and other adults help perpetuate these myths and of course our children believe them. The most damaging thing is that if you believe you were born either good or bad at math then you are limiting your ability to be successful at math. This combined with dyscalculia can make it hard to believe that through effort or practice one can do math as well as everyone else. But if anxiety is faced head on and denied by the significant adults in a child's life, they can in fact do math as well as their peers who do not have dyscalculia. But they all come down to feelings of stress that trigger the fight or flight reaction in us, controlled by the amygdala.

The role of the school, home, and therapists in supporting math anxiety

So, what can we do as adults to decrease the probability of a student developing math anxiety? If we follow the research, we can start by addressing our own attitudes and beliefs about math and the anxiety we feel and the impact it has on the students. As noted earlier, research suggests that students whose parents or teachers are anxious about math not only teach math in a less than confident way providing no confidence that the child is doing the procedures accurately. This in turn makes the student fearful of doing the math in case it is wrong and so the spiral begins. We can address this by being aware that our fears can be noticed by the students and morph into real anxiety.

A quick note about one of the most anxiety inducing events that happen in our school frequently- timed math quizzes, the 20 facts in 60 seconds tests that aim to help students gain automaticity of the facts. The research is not conclusive on whether timed tests are good or bad, there is general consensus that fluency is an important goal for students as automaticity of math facts makes the completion of more complex math problems easier to do as working memory is not taken up by having to complete basic math facts. It is not the fact that the test is timed, that is a factor but more that I am competing against my peers, and everyone will know that I am slower. If the tests are made low stakes, I am only trying to improve my score and no one knows what my score is, then anxiety should be lowered.

Dealing with math anxiety head on

As you can see, math anxiety has an impact on an individual's day to day learning and can affect long term life as an adult if the student avoids careers that involve math. Math is present not just in math classrooms but also heavily in the sciences and other subjects. Negative comments around mathematics are common in today's world unlike statements around reading and literacy. The adults in a child's life can reinforce the anxiety through their own anxiety being communicated verbally or behaviorally, peers in classrooms making statements of negativity around math, choosing that as the class that is skipped or avoided, making light of poorer grades in math as compared to other classes.

As you can see math anxiety is not due to math not being important but rather because it is so important that the child does not want to mess it up. One does not feel anxiety about unimportant things, quite the opposite. So, a good action to take is to reinforce the importance of math and the adults are there to help. A 'little bit of worry' can lead to an enhanced focus on the topic and better performance. Adults can help with this by finding examples of individuals the child looks up to who are good at math, these could be social media influencers, movie characters, etc. Think of movie characters who are good at the sciences - Gru in the Despicable Me series from Illumination Entertainment, the various characters such as Iron Man and Bruce Banner/the Hulk in Marvel series.

Rather than addressing the anxiety indirectly via improving confidence, show students that they are making progress in mastering concepts with a simple graph. It will speed up the process if, as researchers suggest, actual intervention strategies are put into place. This includes teaching students to:

1. Do positive self-talk
2. Deep breathing exercises
3. There are classic children's books that share how their characters deal with math anxiety such as:
    a. Math Curse by Jon Scieszka
    b. When Sophie Thinks She Can't by Molly Bang
    c. The Monster Who Did My Math by Danny Shnitzlein
4. Break problems into manageable steps to prevent feelings of being overwhelmed
5. Provide games and puzzles as this can decrease negative math experiences and increase "I can" feelings
6. Make sure to point out that the child did math work that was

one step easier and accomplished it, this pointing out of previous knowledge and successes enables the student to realize they can do this new work as they have successfully done similar problems earlier

7. This is also not a time to have the student guess, instead show them how to work the problem or do the math facts with them initially and then gradually have them answer with you and then on their own. Watch this clip of me doing an errorless form of learning a math fact and solving a math problem.

8. Being aware of their anxiety by describing how they feel about math and what they feel when doing math. For younger students they could draw pictures or write stories. Older students could write about their emotions or develop a story. This enables them to regulate or remove it to counter their reluctance to engage in math activities.

9. If you think that timed tests are the way to improve math fact recall, please think again, and consider using timed tests only with the child trying to improve their own time and there is no sharing of the time scores with other children. Just as important, call them practice sessions, just like athletes do to improve their performance or time and they are not graded. These should be running records and simply graphed so the child can see the improvement in their performance.

10. Consider how you have a child answer in public, these students are going to need more time to respond, and you will want a backup plan that does not embarrass the child if they suddenly cannot answer the question!

11. Use as many real-world examples and examples from their lives if possible as it makes math relevant and helps the child be vested in learning.

12. Students with dyscalculia also do well working in small groups to complete work as they have someone to collaborate with, hear different thinking or explain how, which helps their own understanding.

13. Professional counseling is an option with guidance counselors in schools, educational therapists, and others able to provide these services.

14. Organize the room to minimize distractions and if it makes sense because of circumstances the child will experience in school or in their careers, gradually increase the distractions until the environment is as normal as possible.

15. Immediate feedback and verbal feedback are preferred as there is nothing worse than waiting for the results of a test or event. Most of us feel some anxiety as we wait for nearly any kind of results.

Chapter Twelve

WHAT TO DO IF YOU SUSPECT YOUR CHILD HAS DYSCALCULIA

Regardless of age, the process, and steps for supporting your child are the same but the focus will be based on their developmental age.

Discovering your child has dyscalculia.

If you are just starting out and have noticed that your child or student is struggling and not hitting the milestones listed earlier your first steps are to:

Talk to someone - if your child is not in school yet, talk to your doctor or someone you trust to listen to your concerns. You can also contact the Child Find Office and ask to speak to someone. Child Find is a free service that every public school district in the United States is required to offer. From birth through age 21, the public-school systems are charged with finding and evaluating those individuals who need special education. It does not matter if your child will be going to a private school or be homeschooled. The services are offered at no charge through the public-school systems, so use them. Go to the website for your local school system and search Child Find!

Of course, if your child has started school, nursery school, preschool, whatever the grade they are currently in, then talk to their teacher or a school administrator to get access to Child Find.

You and your child may want to take a screener to see if there are any indicators that the issue could possibly be dyscalculia. Dyscalculia screeners are not as easy to find as dyslexia screeners but there are options, some with costs and some free.

Once you have completed the screener there will be one of three recommended courses of action the parent or educator can take to help the student. They are as follows:

1. If there are indicators of potential dyscalculia speak with your child's teacher and share your concerns.

Ask for more information on how to have your child evaluated for a learning concern based upon the policies of your public school district, independent or parochial school. If you are a home-school parent, you will want to talk to the home-school liaison from the district or directly approach an independent school psychologist, psychiatrist, or pediatrician.

Maintain your child's self-esteem by reassuring them that they are smart, that anyone can have difficulties with math, and that you are working with experts to determine why and how to support him/her best. If a formal diagnostic procedure is warranted and comes back as positive, share the proper name of the disorder- Specific Learning Disability- Math or Dyscalculia- with your child. There are support groups that can help with the best way to explain it based on your child's age. Explain that there are supports and interventions that can help them learn math, and you are there to help, as are their teacher and/or parents. Plan to use some of the research-based evidence-supported interventions shared later and be sure to share with your child's teacher or parent, as the case will be immediate. Regardless of whether the student is formally diagnosed and provided with an IEP, the process can take a while, and interventions that can make learning math easier should not wait.

2. If there are some risk indicators present but not enough to clearly signal the need for further evaluation, you may want to test again in 6

months to a year but to provide intervention support immediately in the areas of math where the student is struggling.

3. If there are no risk indicators at that time but the parents or teachers see that the student is struggling with math, but there is no clear indication that it is in the areas that will be used for the diagnosis, the report will recommend testing again in a year or two, utilize working memory interventions, talk with the teacher, and focus on early numeracy skills until there is no visible struggle. During the screening and potential diagnostic process, there is the opportunity to have the student be supported via the Response to Intervention process. This can also be implemented regardless of the DySc screener results as it is usually a required step used in the larger diagnostic process to collect data from any interventions used and is a good framework for delivering any interventions that are implemented before a formal diagnosis. In addition, the RtI (Response to Intervention) approach can have benefits to students struggling in math, particularly at risk of dyscalculia and may actually provide information before another screener is done to confirm the presence of indicators of risk (IES, 2021; Berkeley, et al., 2020; Riley-Tillman, 2020). The Response to Intervention (RTI) process is a multi-tiered approach to providing academic and behavioral support to students in schools. It involves a collaborative effort between teachers, parents, and other educational professionals to identify students who are struggling, provide them 58 with interventions and support, and monitor their progress over time. Typically, it involves the following steps: (US Department of Education, 2017) which are: Tier 1: In the first tier, all students receive high-quality, evidence-based instruction and support in the general education classroom. Teachers monitor student progress and may provide additional support or interventions to students who are struggling. Tier 2: In the second tier, students who are not making adequate progress with Tier 1 instruction receive additional targeted support and interventions in small group settings. Progress is closely monitored, and interventions may be adjusted based on student response. Tier 3: In the third tier, students who

continue to struggle despite Tier 1 and Tier 2 interventions receive more intensive, individualized support. This may include one-on-one instruction, specialized interventions, and/or referrals for special education services.

After taking any screener but especially with TouchMath's DySc screener, there is a Recommended Intervention Plan. This section of the report will provide specific recommendations for interventions that will help the student improve the targeted math skills of number sense, math facts, math calculation, and mathematical reasoning. Recommendations for specific teaching strategies, technology tools, or specialized instruction will be grouped according to the needs of the child, the four math areas of struggle, the most effective instructional strategies for students struggling with math, and the most effective learning strategies that a child can use as they master the math or approach mastery (Kim et al., 2022).

Although dyscalculia does not have a cure, there is an ever-growing body of research that supports the effectiveness of using early interventions as soon as a child shows they are having difficulties mastering skills and concepts to improve math (Abd Halim, 2018; APA, 2022; Chodura et al., 2015; Dennis et al., 2016; Mahmud, 2020; Menon et al., 2021; Mononen, 2014). Research also supports matching the intervention to specific content areas of math to increase the intervention's effectiveness and personalize it to the child (Chodura et al., 2015; Haberstroh & Schulte-Korne, 2019; Nelson et al., 2022). However, rather than wait for a child to fail or prove they have a disability, interventions should be implemented immediately as the procedures for diagnosis can take months or longer while the child loses time in successfully building the foundational skills and concepts needed. It has been shown that with evidence-based interventions, students with dyscalculia can achieve age-appropriate math levels (Bailey et al., 2020; Dennis et al., 2016; Kuhl et al., 2021).

It must be noted that most intervention research has been done with elementary-age students rather than secondary students. We have assembled the key intervention strategies supported by the evidence, which can be readily done by parents and educators, have a strong to moderate effect size in the research and, where warranted, targeted them to the particular area of struggle for the child. The interventions we have focused on are not all-inclusive but they are effective in all areas of math. Because of that we have not felt the need to go deeply into each intervention but rather to give enough of a description for practitioners and researchers to recognize the intervention. We have defined interventions for the struggling student as having four distinct components. The teacher/educator determines the best combination to implement based on the students' needs at a particular time. The four areas are:

1. The Child-student supports that impact the ability to learn math.
2. The Math-the particular area of math that will be targeted.
3. The Instructional Strategies- the strategies the teacher chooses to use.
4. The Learner Strategies- strategies the teacher has the child learn as temporary or permanent supports.

The interventions are presented in a specific order deliberately. It is critical to support the child first, in order to prevent further anxiety or disengagement with a task that they find difficult and may cause them to believe they are 'dumb' or incapable of mastering math.

Chapter Thirteen

One thing to remember is that your plan will not be for just the first years of your child's life but at the very least until they launch into adulthood. Your continuing support is critical to your child's ability to accept themselves as an individual with a disability and adapt to it. Dyscalculia may be a disability but just like individuals with other disabilities such as near sightedness, deafness, or dyslexia, it should not be an impediment to your child growing up to have the career of their dreams and be a totally self-sufficient adult.

So now that we know some more about how math develops in all of us, that all of us have a variety of feelings, anxieties, and memories about how we learned math and interact with it daily and that there are easy things we can do at home let's pull all of this together in a plan.

Firstly, a plan is something that can be quickly and easily done and does not have to be overly involved. We can start with accepting our child or student as someone who will need to make an extra effort and need extra time to meet grade-level expectations, and that knowledge the student has of being different can add to the anxiety. It is reasonable to believe that since the anxiety impacts learning it should be factored into instructional design for these students. Add to this comorbidity of dyslexia, ADHD, math anxiety we discussed earlier, and a student's engagement in learning is most likely impacted. It will be important for teachers and parents to provide affirmation and learning environments that make engagement and success more likely.

Let us take the plan we need to develop and turn it into 3 phases- the early years (3-10 years of age), the middle years (ages -11-16 or middle school/early high school) and transition to adulthood (17+). Each stage will require your unwavering support, but each has differences.

Building a plan to support the child – The early years

If your child is between 3 and 5, just starting preschool or K, and you are not seeing them hit some of the milestones shared earlier in this chapter your plan would definitely start with not worrying but rather making sure to add some of the number sense activities to your daily time with your child and have fun with them. This is not a time to become anxious and in turn inadvertently communicate that worry to your child. So, your plan might look something like this.

1. Have fun with math with your child.

2. Talk with your preschool or kindergarten teacher or administrator about your concerns or call the local Child Find number

3. Use a screener and based on the results talk to the school and proceed to diagnosis

4. If your child is struggling the best thing to do is begin interventions immediately as it is easier the younger the child and this is like dyslexia- it will not go away but it can be a lot easier for your child. Maintain a positive attitude and reassure your child that you are there to help, there is nothing wrong with them and you will work to figure it out.

5. As your child moves through elementary into middle and high school keeps the positive attitude, attend IEP and 504 meetings, and absolutely advocate for your child. They have a disability and under the law are entitled to support so that they can get the same education as their peers not struggling with math. During the IEP meeting there are a number of accommodations that can be discussed for your child depending upon the severity of the

dyscalculia. And just because they need a certain support to start does not mean they will always need it. The opposite holds true also, what they do not need at the age of 7 may be needed at the age of 16 and as they enter the world of post-high education.

6. Look for additional therapeutic support from local educational therapists. They can be found via contacting the AET or ALTA or I have listed some resources at the back of the book. They usually charge but if money is an issue ask if they have a sliding scale. Oftentimes teachers in the district can help also with additional tutoring.

7. Look at joining any number of organizations that are made up of parents, individuals with dyscalculia and other learning disabilities, and learn more. One of the best organized is the Learning Disabilities Association of America, https://www.ldaamerica.org.

Building a plan to support the child – The middle years

If your child is not screened or diagnosed until after Grade 3, there is still much that can and must be done. Because a child is born with dyscalculia it has been there, but they may have been able to keep up just enough in school that it was not readily apparent, or it is not as severe.

If your child's teacher or you notice that math is a struggle, definitely control your anxiety and seek advice. The same goes for the advice I gave to those with younger children - this is not the time to become anxious and in turn inadvertently communicate that worry to your child. So, your plan might look something like this.

1. Have fun with math with your child – do activities such as cooking, grocery shopping etc. that are more age appropriate and please check out suggestions in Chapter 6

2. Use a screener and based on the results talk to the school and proceed to diagnosis

3. Talk with your teacher or administrator about your concerns and consider requesting that they place your child in an intervention program for a period of 6-12 weeks and then meet to decide if there needs to be a formal move to a Child Study and a formal assessment process begun.

4. Look at joining any number of organizations that are made up of parents, individuals with dyscalculia and other learning disabilities, and learn more. One of the best organized is the Learning Disabilities Association of America. ldaamerica.org

5. If your child is struggling the best thing to do is begin interventions immediately as it is easier the younger the child and this is like dyslexia - it will not go away but it can be a lot easier for your child. Maintain a positive attitude and reassure your child that you are there to help, there is nothing wrong with them and you will work to figure it out.

6. As your child moves through the later years of elementary into middle and high school keep the positive attitude, attend IEP meetings, absolutely advocate for your child. They have a disability and under the law are entitled to support so that they can get the same education as their peers not struggling with math.

7. Begin to teach your child how to advocate for themselves- to share that they have a disability and need a little more time, but they can do the math. No one thinks of wearing glasses or having dyslexia as that much of a big deal anymore and dyscalculia is the same.

8. Teach your child positive self-talk and explain to them they may have a disability but between them, family, and school you will figure it out.

9. Search for stories of individuals with math struggles and dyscalculia who grew up to be what they dreamed of. This includes actors, scientists, etc.

10. Look for additional therapeutic support from local educational therapists. They can be found via contacting the AET or ALTA or I have listed some resources at the back of the book. They usually charge but if money is an issue ask if they have a sliding scale. Oftentimes teachers in the district can help also with additional tutoring. And there are schools who specialize in working with students with learning disabilities and also colleges. See the back of the book and the TouchMath website for more information.

Building a plan to support the child – Transitioning to adulthood

And yes, sometimes the diagnosis does not happen until your child is well into middle or high school but there is still much that can be done, and dreams do not need to be discarded. It may just take longer, and more support be needed. So everything that is in the previous section and the following.

If your child is not screened or diagnosed until after Grade 6 there is still much that can and must be done. Because a child is born with dyscalculia it has been there, but they may have been able to keep up just enough in school that it was not readily apparent, or it is not as severe.

If your child's teacher or you notice that math is a struggle, definitely control your anxiety and seek advice. The same goes for the advice I gave to those with younger children - this is not the time to become anxious and in turn inadvertently communicate that worry to your child. So, your plan might look something like this.

1. Use a screener and based on the results talk to the school.

2. In secondary school or earlier, as you explore educational opportunities for after high school, make sure to ask about what the college, university or trade school does to accommodate your child's learning needs. Ask about the Disabilities Services

department which is responsible for the observation of the Americans with Disabilities Act (ADA) and the continued monitoring of Section 504 in college. Please note that the responsibility rests primarily on the shoulders of the individual to seek the services and they will primarily consist of the institution providing the accommodations needed to ensure the student can access their classes and materials as well as demonstrate mastery of the coursework.

3. Have your child actually advocate for themselves - to share that they have a disability and need a little more time, but they can do the math.

4. Search for stories of individuals with math struggles and dyscalculia who grew up to be what they dreamed of. This includes actors, scientists, etc.

5. And there are schools who specialize in working with students with learning disabilities and also colleges, this is a great time to start college or trade school visits and reinforce with your child how to advocate for themselves. See the back of the book and the TouchMath website for more information.

6. Looking at joining any number of organizations that are made up of parents, individuals with dyscalculia and other learning disabilities and learn more. One of the best organized is the Learning Disabilities Association of America. ldaamerica.org.

What to know and ask when building the IEP or 504 plan

If your child is diagnosed with dyscalculia there are 3 possible outcomes:

1. They need an IEP, the individualized education program, which lays out how the schools your child attends will provide them with the additional educational supports they need. The IEP is not necessarily permanent, it can be needed at various stages of

life and until your child can advocate for themselves that is an important part of the support you must provide. Once your child is an adult, they are entitled to protections in the U.S. under IDEA and you can learn more about IEPs, 504 plans, and the processes involved by going to your school district website or just googling IEP.

2. They do not have a severe enough educational discrepancy to warrant an IEP, a 504 will provide additional support other than specialized instruction so that your child can learn alongside their peers.

3. They are proceeding well enough with their math that they currently do not need an IEP or 504. In this case your awareness of your child's progress and talking with their teachers to ensure that if things begin to fall behind, a meeting can be pulled together and a support plan, 504, or IEP can be quickly put in place and minimal educational time and frustration happens.

As this process unfolds you will find yourself full of questions so I have gathered some of the most common ones asked and the ones you should probably be asking as well.

Common Accommodations for Dyscalculia

There are a number of common accommodations that can be put in place to support your child and make it possible for them to access the same curriculum as their peers. This means they will be expected to meet the same state standards as their peers and take the same state test. It is important to keep in mind that not all of the accommodations may be needed, and they may change over time.

The school psychologist or diagnostician, as well as the Special Education team members who are present at the IEP or 504 meeting, should be aware of these and be able to explain which are most appropriate for the plan being developed. They can also shed light as to how long they may be needed. The typical accommodations include:

- Calculators. Unless it substantially alters the requirements of the course or computation is the focus of the lesson - you will need to ask what is permissible
- Additional time to complete assignments and assessments including state and local tests.
- Breaks during exams
- Quiet space to do work when reduced distractions are needed
- Note-taking accommodations such as the ability to record lectures or share notes with another student or access the teachers notes
- Visual aids such as lines drawn on problems to separate ones, tens, and hundreds; columns; manipulatives; or graph paper to help solve problems
- Reference sheets such as addition and multiplication tables
- Formula sheets
- Text to voice tools
- Tutoring, homework assistance, monitors who can point out errors and provide additional opportunities to attempt the work
- Worked examples
- Different test designs
- Small group or 1:1 learning and performance opportunities
- Asking questions that have the student showing how they are thinking
    - Where would you go for help if you were having a problem

solving it?

- Can you tell me where you made the mistake?
- Is this always the way to solve this type of problem or are there other ways?

In Summary

When putting the plan in place there are some key things to do that will help you and them. They include:

- Support your child or student. There is so much you can do as a parent or teacher and when working together, there is nothing you cannot do.
- Talk honestly with your child based on their ability to understand dyscalculia and that they will learn math with more difficulty and need more time. Get support or advice if you feel the need.
- Make sure all of the adults in your child's life understand, put this in the context of dyslexia or needing eyeglasses as those disabilities at one time were not understood and there were misconceptions.
- Put math in the real world as that is what it is for, to solve real-world problems. How do I share 3 cookies evenly when there are 4 of us?
- Be positive. Your attitude and enjoyment of being with them translates in so many ways into how they view themselves and their ability to do math, but the research has shown it actually has an impact on their achievement in math.
- Get support for yourself, there are support groups out there for parents of students with dyscalculia and all options should be explored. Some of the more prominent ones are listed in the resources section.

- Start early, as soon as you see there are struggles, do something – talk to the teacher, act. Preschool and transitional kindergarten should be definite if your child is still young. Your support at home, using interventions and tutoring and working the Child Study Team (CST) at school should be discussed.

- Focus on the math areas of number sense, operations, and reasoning. These are the foundations of nearly all math, and we all know the importance of a strong foundation for everything from a new building to new knowledge. Including spatial reasoning may also make sense based on the current research.

- Focus on those executive function skills of paying attention, thinking things through. They go hand in hand with math and as executive function abilities grow so do math abilities.

- Games and computer assisted instruction should not be avoided, just not overused. Humans teach math, computers do not.

PART V

In Part V you will learn:

- What makes an intervention effective for a student with dyscalculia?
- What interventions are currently supported as being effective for the child and the math domains impacted
- The instructional and learner strategies that are important to implement
- How to implement the interventions
- And activities to do that combine these

Chapter Fourteen

INTERVENTIONS

Just what is an intervention?

The Merriam- Webster Dictionary defines intervention as "the act of interfering with the outcome or course especially of a condition or process (as to prevent harm or improve functioning)." The interventions are not done instead of the regular education but rather at the same time - they provide more time and support in math so the student can participate in the regular classroom math period. They are a series of actions that an educator takes to help a student who is struggling, either academically or behaviorally. The goal is for the intervention to be effective and the intervention to end, to close the achievement gap between students with dyscalculia and typical students.

Regardless of the content that is being taught there are a number of common components found in effective interventions for students who are struggling, and which apply to students with dyscalculia (Chodura, et al., 2015; Fuchs, et al., 2008; Melby-Lervag, et al., 2016; Mononen, et al., 2014; Zakariya, 2022). These are:

- The earlier in a child's academic path the intervention can be done the more effective it is as the neural pathways and regions of the brain are more flexible the younger the child.

- Intervention is delivered explicitly and systematically so that math concepts and procedures are modeled in multiple ways- the adult modeling how, then working alongside the student and gradually releasing support so that the student can practice on their own.

- The intervention is used to actually teach and apply the math skills and procedures that are being targeted. Just practicing math tables is not effective unless the child can use those facts to solve real problems.

- Feedback is immediate and verbal as hearing the feedback is more effective than a written grade or comment. Verbal feedback allows the adult to elaborate or provide examples as needed.

- The intervention is individualized for that student. Data from that student's responses are used to adjust instruction immediately, this includes adjustments in intensity which can include class or small group size, location of the setting- pull out or in the general education classroom, days per week and length of the intervention sessions.

- Intervention is comprehensive and targets all the aspects of dyscalculia and any comorbid disabilities so that math anxiety, dyslexia, ADHD, as well as the math are addressed comprehensively and not in an isolated manner.

- The goal of the intervention is to have the student receive primary instruction in the core curriculum with the intervention being supplemental.

For this to be effective, educators use research-based and evidence-proven interventions as soon as possible, the earlier the better, while children are still developing the basic foundations for numeracy. "Early maths matters: Those who start behind stay behind – and the gap widens in primary school." (Education Endowment Foundation, 2020) Interventions can include remediations as there may need to be certain skills retaught. But they also include accelerating the learning process in order for the child to catch up. The terms intervention and remediation are commonly used interchangeably in education, but the definition of remediation is to correct something that is wrong to close a gap so that instruction can continue. Because these students are typically already behind, remediation does nothing to bring them to grade level, it only keeps them from falling further behind. An intervention has to be done at the same time as regular instruction is happening so that the child does not fall behind, it needs to accelerate the process so the child can catch up and stay up.

As stated earlier in this book, an achievement gap has serious consequences for students with dyscalculia: lower rates of high school graduation, lowered career expectations resulting in lower lifetime earnings, higher levels of unemployment and higher rates of depression. Intervention using effective tools and instructional strategies, as early as preschool and kindergarten, increases the probability that those with dyscalculia can close the achievement gap with their typically developing peers and counter the results of lowered academic achievement and expectations.

Research continues to support the need for early screening, identification, and intervention, The rapid growth of the brain and its extreme malleability during the windows of opportunity in the early years of instruction make the time from preschool to 3rd grade a critical intervention period for math development (Dehaene, 2020.) Screening for difficulties with number sense, math facts and calculations, and mathematical reasoning and immediate intervention is more effective in younger children and increases the chances of sparing a child the consequences of poor numeracy skills and will likely have significant implications for their future academic and career success. The earlier the identification and appropriate intervention, the easier it is to remediate; free or inexpensive screening measures can identify at-risk children as young as 3 in the more severe cases and through adulthood.

Keeping the above in mind, it is essential to understand that the longer we wait, the more difficult the child and teacher's path and the higher the probability of math difficulties and the consequences of those difficulties.

We have assembled the key intervention strategies supported by the evidence, which can be readily done by parents and educators. These have a strong to moderate effect size in the research and, where warranted, can be targeted to the area of struggle for the child. The interventions focused on here are not all-inclusive. But they are effective in all areas of math content and in the case of the instructional and learner strategies in every content area. Because of that we have not felt the need to go deeply into each intervention but rather to give enough of a description for practitioners to recognize the intervention and because of their pedagogical knowledge immediately implement it correctly. We have defined interventions for the struggling student as having four distinct components. The educator determines the best combination to implement based on the students' needs at a particular time. The four areas are:

1. **The Child –** student supports that impact the ability to learn math.

2. **The Math –** the area of math that will be targeted.

3. **The Instructional Strategies –** the strategies the teacher chooses to use.

4. **The Learner Strategies –** strategies the teacher has the child learn as temporary or permanent support.

The interventions are presented in a specific order deliberately. It is critical to support the child first, to prevent further anxiety or disengagement with a task that they find difficult and may cause them to believe they are 'dumb' or incapable of mastering math. The next step is to determine what is the specific math skill, concept or procedure that is being targeted. From there one can select the right instructional strategies to use and the right learner strategies to have the student use.

Chapter Fifteen

THE CHILD - STUDENT SUPPORT INTERVENTIONS

This group of interventions is aimed at supporting the student by addressing math anxiety, self-efficacy, working memory, and attention span. If the child is struggling with these, the effectiveness of any intervention done around the actual remediation or acquisition of math content will be impaired. It is difficult to focus on learning how to parallel park a car the first time if fire engines are going by, music is playing, you have a major headache, and folks are asking you if you want to go and play.

Math anxiety interventions

Math anxiety is a common experience for many students, and it can significantly impact their academic performance and confidence. Fortunately, there are a number of interventions that can help to reduce math anxiety (Jordan et al., 2013). These include:

- Positive self-talk: Encourage students to use positive self-talk, such as "I can do this," "I am good at math," or "I can learn from my mistakes." Positive self-talk can help to boost confidence and reduce anxiety.

- Breaking down complex problems: Encourage students to break down complex math problems into smaller, more manageable steps. This can help to reduce feelings of being overwhelmed and make the problem-solving process less daunting.

- Math games and puzzles: Incorporating math games and puzzles into instruction can help to make math more enjoyable and engaging. This may help to reduce negative associations with math and reduce anxiety.

- Errorless learning: Errorless learning is an instructional strategy that increases the likelihood that the student always responds correctly. Because they don't have to guess and get it wrong possibly, anxiety is lowered. Rather than respond to an incorrect answer from a student with "no, the answer is not 10. 7+4 =11"; the teacher repeats the problem with the correct answer so that what is retained in memory is the correct response. "7+4 = 11, say it with me." Initial mistakes are prevented so that they are not the most recent memory of the student.

- Activating prior knowledge is a strategy that alerts the student that they will be encountering new information but that there are previous links currently in their memory that are connected. Teacher prompts might include, "tomorrow we are going to be learning double-digit addition, very much like what you already do, but there will be two numbers instead of one." Knowing I have already learned this means it is not new; I can review it and remind myself or relearn it.

- Real-world applications: Show students how math is used in real-world situations. This can help to make math more relevant and meaningful and reduce feelings of anxiety.

- Relaxation techniques: Teach students relaxation techniques, such as deep breathing or visualization, which can help to reduce anxiety in math class or during math tests.

- Peer tutoring or support groups: Encourage students to work with peers who are supportive and can provide guidance and feedback. This can help to reduce feelings of isolation and increase confidence in math. And having a peer explain it in a different way can make the difference in learning it as it a student, a peer explaining rather than a teacher.

- Professional counseling: If a student is experiencing severe math anxiety impacting their academic performance, it may be helpful to refer them to a professional counselor who can provide additional support and guidance.

- Classroom setup: allowing the student to have a quiet area or one where they are not distracted and can use any special supports that may signal to peers their 'differentness' also reduces anxiety.

- Organization skills that prevent failing to turn in an assignment or being late also reduce anxiety.

Reducing math anxiety is not a quick fix, but a process that takes time and effort. It is important to be patient and persistent in implementing interventions and providing ongoing support and encouragement to students who are struggling with math anxiety.

Self-efficacy interventions

Self-efficacy has long been identified as a major contributor to success or lack of success in math. Students with high self-efficacy, the belief that they have done well in the past, an 'I can do it' attitude, and a belief they can do the presented task help determine a student's engagement with the math task (Bandura, 2012). It can determine how much effort they will put into the lesson and how long they will tolerate frustration with the task's difficulty. If they are successful, it reinforces the can-do attitude and their judgment of their own competence, and vice versa with failure. If my self-efficacy is high, my math anxiety will be low, and vice versa. Moreover, this also increases mastery of the targeted math skill or concept (Schaeffer, et al., 2015) and is a strategy that can be combined with any instructional, learning, or student support strategy. It can also be provided directly to the student so that their self-efficacy improves. The most effective ways to help bolster self-efficacy are:

- Ensuring the students participate in and master a task via errorless learning (see above).

- Giving verbal, not written feedback from others including peers, teachers, parents, and more proficient adults with verbalization by the adult of the student's capability being the most important to support self-efficacy (Bandura, 2012; Zakariya, 2022).

- Providing the emotional aspect of feeling secure and not fatigued or stressed also bolsters feelings of competence and self-efficacy.

- Providing the child with anxiety coping strategies such as breathing, and relaxation techniques can also address this issue and are particularly useful for high-stress situations such as test-taking. As you can see, some strategies address multiple issues.

- Gamification or computer-assisted instruction also addresses self-efficacy and anxiety issues as the child competes against themselves, has immediate feedback, and is in a low-risk non-social environment (Baten & Desoete, 2018; Kohn et al., 2020).

Working memory interventions

- Students with dyscalculia commonly have issues with working memory. Working memory is an important cognitive skill that allows us to temporarily store and manipulate information such as the retrieval of math facts. It is considered to be a domain general area of struggle (Geary, 2013). It plays a crucial role in a wide range of academic activities, including reading, math, and problem-solving. Improving working memory can, therefore, have a significant impact on academic performance. Effective interventions include:

- Presenting information in chunks that can be more easily accessed and maintained when completing problems. Chunking involves breaking down information into smaller, more manageable pieces of information (Dueker, 2022). This includes teaching the student to remember the number in chunks; to break down longer strings of numerals into smaller pieces, think memorizing a phone number, or using rhythmic skip counting. Other examples include grouping vocabulary by word families - such as the operations and turning them into an anagram or silly phrase or using a modified Frayer chart.

- Another strategy is taking multistep tasks and presenting the information in subsets of the task, making sentences shorter, or using paper to reveal information slowly. Drawing lines where place value is being worked or circling key information in word problems are all simple steps to take that help the student, and students can be taught to do them when performing other tasks.

- Providing wait times that are long enough to process info, especially new concepts, skills, and procedures is an effective intervention (Geary, 2013; Gersten, 2011).

- Using mnemonic strategies that involve using mental images or associations to help students remember information. Teaching students to remember a list of words by creating a story or visualizing each word in a unique way.

- Doing a "see, say" game with numbers. I do not show the student the number but rather say it and the student must write it. This is a great way to have fun and no pressure while practicing transcribing and working to increase short term memory skills. It can also be reversed, and I show the student a number, cover it up and they must say and /or write it.

- Verbal rehearsal, which involves repeating information aloud to oneself to help store it in working memory is another strategy (Bergman-Nutley & Klingberg, 2014).

- Visual-spatial strategies involve using mental images or spatial relationships to help remember information (Arsalidou & Taylor, 2011). Teaching students to use a spatial map to remember the location of items in a room is an example.

- Computer-based games and training programs are available that can help improve working memory. These programs often involve games or activities that challenge students to remember and manipulate information.

When implementing any of these interventions, it is important to provide explicit instruction (more on that in a couple of pages) and practice opportunities. Additionally, it is important to ensure that the interventions are appropriate for the student's age and cognitive abilities and that there is proof of efficacy for that intervention.

Attention Span Interventions

Attention span is a critical factor in a child's ability to learn and perform well academically. Children with shorter attention spans may struggle to stay focused in class or complete assignments. For children who have been diagnosed with ADHD, a consultation with the educator or therapist is a needed step. Intervention strategies that can be used to increase attention span in children include:

- Break tasks into small, manageable chunks: Smaller, more manageable chunks of information or tasks make them less overwhelming and easier for a child to accomplish. Completing the first 3 steps is always easier than doing a set of 10 steps. This can help children stay focused and engaged.

- Increase physical activity: Children benefit from regular physical activity, which can help improve focus and concentration. Encourage activities such as recess, physical education classes, or after-school sports programs. Children benefit from short movement breaks during the day. This can help release energy and increase focus during other activities. Another example would be the use of stations or centers in the classroom along with scheduled rotation. This builds in movement during the class period and also provides structured routines. Even when we are training, the best practice is to change the activity or focus every 15-20 minutes.

- Use visual aids: Visual aids, such as diagrams, charts, or videos, can capture a child's attention and make the learning experience more engaging and memorable. Another idea is to use the student's interest to inspire the visual supports. We often use the student's particular interest or "fascination" to help us develop the visual supports. The student is more likely to pay attention to something that normally he/she might ignore.

- Provide clear instructions: Children with shorter attention spans benefit from clear, concise instructions. Break down tasks into steps and give frequent reminders about what needs to be done.

- Use positive reinforcement: Praise and rewards for good behavior and focus can be motivating and help children stay on task.

- Include errorless learning: Always start a problem-solving response by the teacher with the correct answer so that students can hear and remember what is correct rather than remembering that the first response by the adult was, wrong, the answer is 8 not 7.

It is important to remember that every child is different, and what works for one child may not work for another. Be patient and persistent in trying different strategies to find what works best for each child. Also, consult with a child's teacher or a mental health professional if the child's attention span is consistently shorter than expected for their age.

Chapter Sixteen

THE MATHEMATICS INTERVENTIONS

Mastery of number sense, being fluent and accurate in arithmetic facts, understanding whole number computation, and mastery of problem solving and reasoning are critical for being able to master the math progressions. These foundational areas are positively related to performance in other math skills and can predict performance in later years on math assessments (Fuchs et al., 2012: Jordan et al., 2013). Students with dyscalculia show evidence of difficulty in these areas and interventions that target these distinct areas of struggle are effective.

The research suggests that the deliberate intervention on or training of specific areas of math weaknesses works by actually enabling the student to learn the needed skill or concept using the typical neural pathways and decreasing the future load on working memory and attention systems once mastered (Aquil, 2020; Chodura et al., 2015; Dennis et al., 2016) This constitutes the strengthening of the typical pathways and areas of the brain rather than compensatory areas being enabled. Plasticity in the neural circuits or adding additional white matter, myelin, to the neural pathways was supported (Menon et al., 2021).

It must be noted that by intervening specifically in these four areas, other areas of math need not be ignored and, in fact, must not be. It is also important to note that by targeting the areas identified as domains for a dyscalculia diagnosis, there will be the potential to impact the ability of the child to show the mathematical habits of mind and skill that the National Council of Teachers of Mathematics (NCTM) feels are needed in order to be successful in mathematics as they grow older.

So, let's look at those 4 areas of math and things you can do:

Number sense

Number sense is a rudimentary system we are born with, but its continued development makes formal math possible. Therefore, intervention in the form of the explicit instruction of number sense is critical to improving this skill when a student shows signs of struggle (Dennis et al., 2016; Geary, 2013; Nelson, 2019), as number sense enables students to succeed in the classroom (Halberda, et al., 2008; Kucian, 2015; Kucian et al., 2011; Wilson, Dahaene et al., 2006).

Instructional interventions that have shown success in improving number sense include (Fyfe, 2019):

- Games that require counting and recognizing that one number is larger than another, such as Chutes and Ladder or TouchMath's Connect 1. This focuses the intervention on attaching the numerical symbol to the magnitude (4>3) and helps with addition and subtraction.

- Using manipulatives and visual aids or the concrete-representational- abstract framework to understand the connection between the physical object and its abstract symbol. Using the TouchMath technique has been shown to be effective in teaching number sense and later the math facts and calculations. It is a multisensory approach, combining visual, auditory, and tactile actions as the student places dots or object pictures on the concrete object, drawing or the abstract numeral. This simultaneously allows the child to connect the concrete, semi-concrete, and abstract version of the numeral. As the child reaches mastery in number recognition, the dots are removed.

- Lay out a set number of manipulatives. This can include dice, counters, TouchMath numerals, texture cards, etc. (Yikmis, 2016.) Have the child count out the number of manipulatives you laid out, making sure to touch each one as they count aloud. They can also select the symbol that represents the number of manipulatives if they are at that stage of matching.

- The use of the TouchMath BiDiWi sheet is an effective instructional strategy. The use of the TouchMath BiDiWi sheet is an effective instructional strategy. Have students place the correct number of items in the Build It box on the BiDiWi. Then have them draw the physical item or represent it with dots, squares, etc. before attaching the appropriate symbol to the drawing in the Write It box.

- Multisensory math instruction involves using a variety of senses to help children learn math concepts, this includes having them touch, look at, hear the numeral, and say the numeral to enable greater connections in the areas of the brain needed for memory of the numerals. For example, using visual aids, such as pictures and diagrams, can help children understand math concepts, while tactile activities, such as tracing numbers, can help children develop their phonological processing skills.

Number sense is an important aspect of math that involves understanding the relationships between numbers and their properties. Learner strategies can help students improve their number sense in math and in some cases, they will continue to use these strategies through adulthood. Examples of these include:

- Use fingers to count with or finger tapping, drawing TouchPoints. A number of the adults I interviewed still finger tap, touching their fingertips together discretely as they remember certain math facts. Some still draw small dots on numerals when they are working on a calculation problem on paper.

- Play math games: Math games can be a fun and effective way to develop number sense. Board games for students 6 and under such as Candyland, Go Fish, and Connect 4, as well as games that use spinners or dice, can reinforce number sense skills while providing for engagement fun. These are actually practice that the learner can continue on their own. As adults, the use of board and virtual games that require scoring, using dice keep number sense well used.

- Use manipulatives: Manipulatives are physical objects that can be used to represent numbers and help students visualize math concepts. Examples of manipulatives include blocks, tiles, number lines, and Touch Numerals. Learners can choose to use manipulatives to solve many problems and in everyday life. Thinking cooking- pure manipulatives and math and a great way to practice fractions and time management

- Estimate and round numbers: Estimating and rounding numbers can help develop a sense of number magnitudes and improve mental math skills. Practice rounding to the nearest ten, hundred, or thousand is an example.

- Compare numbers: Comparing numbers develops a sense of the relationships between numbers. Practice comparing numbers using symbols such as <, >, and =.

- Break numbers down: Breaking numbers down into smaller parts to understand their properties and relationships. For example, break 24 into 20 and 4 or 10 and 14 to help understand the composition of the numeral.

- Visualize numbers: Visualizing numbers such as picturing numbers on a number line or using diagrams.

- Practice mental math: Practicing mental math also improves number sense and the ability to do calculations mentally. Practicing basic operations such as addition and subtraction, then progressing to more complex problems is the recommended progression.

- Counting games: Counting games can help children practice their phonological processing skills by counting aloud. Examples of counting games include counting objects in a room, counting steps as the student walks up and down stairs, and counting the number of times one can bounce a ball.

- Math manipulatives: Blocks, beads, and counters, help children develop their phonological processing skills by allowing them to physically manipulate objects while orally counting and doing 1:1 correspondence problem.

- Music activities: Music-based math activities can help children develop their phonological processing skills while having fun. Examples of music-based math activities include singing math songs, clapping out math problems, and playing math games that involve music. Have you ever noticed that most of us use rhythm or musicality when we recite long strings of numerals.

Memorization of arithmetic facts

Chodura et al. (2015) and Dennis et al. (2016) found in their meta-analysis of interventions for children with mathematical difficulties that training in basic arithmetical competencies, the math facts, was especially effective for the child with dyscalculia.

Instructional strategies that have shown success in improving memorizing math facts include:

- Using the TouchMath technique has been shown to be effective in teaching math facts. The multisensory approach, combining visual, auditory, and tactile actions as the student places dots or object pictures on the concrete object, drawing or the abstract numeral; simultaneously allows the child to connect the concrete, semi-concrete and abstract version of the math fact. As accuracy and mastery are met, the dots are removed (Fletcher, et al., 2010).

- Extensive rehearsal or regular practice. Setting aside regular practice time each day to help the child build fluency with arithmetic facts. This could involve using flashcards, practicing mental math, or playing math games that focus on arithmetic skills such as dominoes. Incorporating games and activities that are fun makes memorizing arithmetic facts more engaging and enjoyable for the child. This could include using online resources, math apps, or board games that focus on arithmetic skills. Memorizing math facts is easy to make into short fun, beat my last score games.

- Use visualization to help the student visualize arithmetic facts in their mind. Encourage them to create mental images or use manipulatives such as blocks or counters to help them visualize the facts.

- Focus on one set of facts at a time to not overwhelm the student by trying to memorize all arithmetic facts at once. Instead, focus on one set of facts at a time, such as addition or subtraction, and gradually add more sets as mastery is reached.

- Provide positive reinforcement, especially adult verbal praise that is specific to the problem or task performed. This can also include a small reward if appropriate when progress or mastery of arithmetic facts is demonstrated.

- Being patient and supportive and remembering that memorizing arithmetic facts can be a challenge for the student with dyscalculia is essential. Encouraging the child to keep practicing and trying their best is critical.

Use of quick sprints or math fluency tests to assist in memorization of math facts is also needed. It is important that these timed tests do not have the student compete against others; only themselves for increasing accuracy and then speed. The research is not conclusive on whether timed tests are good or bad, there is consensus that fluency is an important goal for students as automaticity of math facts makes the completion of more complex math problems easier to do as working memory is not taken up by having to complete basic math facts.

Learner strategy interventions that are especially effective for memorizing math facts include:

- The calculator is probably the most used of learner strategies for when a math fact cannot be recalled. The one downside is that if the numbers are entered incorrectly, then of course the answer is incorrect. The solution is to perform the calculation 3 times and to say the problem as it is entered as there appears to be the ability to recognize that what one says is not the same as what one sees.

- Fact tables that the child fills in only for the facts not memorized. Having a simple facts table accessible is a strategy that can be used throughout life. The most common one is the hundreds chart which has all the multiplication facts for 1-10.

- The use of counting all or on with one's fingers, placing TouchPoints on numerals or tally marks to support counting. This is the same technique used in number sense.

- Practicing the facts over and over again can help to reinforce them in the memory.

- Remembering the patterns and chunking long numbers such as the sing-song cadence of saying the 5's tables and chunking phone numbers.

- Number rhymes can help children learn number sequences

and improve their phonological processing skills. For example, reciting "One, two, buckle my shoe" can help children learn the numbers one to ten in sequence. This same process is useful as the individual gets older, think reciting the 5's tables, we automatically attach a singsong rhythm to them.

- Using the Frayer-model, BIDIWI, and other visual aids, such as charts, diagrams, or pictures, can help to make math facts more concrete and easier to remember.

- Finding and playing games involving math, such as "Math Bingo" or "Math War" to boost motivation.

Whole numbers computation that is accurate and fluent

Understanding and being fluent in using whole numbers has been positively related to achievement in other math skills and is predictive of subsequent algebraic reasoning success. These skills are the foundation for many of the subsequent math skills needed for the entirety of the math progression, and failure to master one can impact future fluency (Avant & Heller, 2011; Calik & Kargin, 2013; Fyfe, 2019; Fuchs et al., 2012; Jordan et al., 2013; Kim et al., 2022). When students learn to do the operations, they are evaluated on accuracy or successful acquisition of the skill and then fluency or the speed at which they can perform the computation. Because of the importance of speed and accuracy or automaticity, interventions for working on whole numbers and their operations should ensure the student can do the operation and then increase the speed or fluency. Research also supports the idea that frustration and anxiety decrease if a student can perform the operation accurately and then move to increase the speed of computation (Fuchs et al., 2008).

Instructional interventions that have shown success in improving math computation are:

- Timed trials or practices with paper and pencil where the student only competes against themselves to improve accuracy and speed.

- Computer games that enable practice for memorization and fluency.

- Using concrete and semi-concrete versions of the facts as well as visual aids is effective in teaching computation. When presented using a multisensory approach these activities have been shown to be effective. In addition, the provision of additional concrete referents such as finger tapping and semi concrete referents such as TouchPoints or tallies helps students move forward in the progressions (Abdou, 2020; Cihak & Faust, 2008; Green, 2009; Simon & Hanrahan, 2004; Urton et al., 2022; Yikmis, 2016.)

- Flash cards and dominoes, with the focus being to increase accuracy of their answers before increasing the speed of the answers they provide.

- Conceptual understanding of arithmetic procedures such as using a mind map or mathematical modeling to assist in conceptual understanding (Powell, Fuchs, et al., 2009).

- Word problem-solving such as breaking problems down into smaller steps (Fuchs, Fuchs, et al. 2012)

Learner strategy interventions that are especially effective for math computation include:

- The calculator is probably the most used of learner strategies for when a math fact cannot be recalled. The one downside is that if the numbers are entered incorrectly, then of course the answer is incorrect. The solution is to perform the calculation 3 times and also to say the problem as it is entered as there appears to be the ability to recognize that what one says is not the same as what one sees.

- Fact tables that the child fills in only for the facts not memorized. Having a simple facts table accessible is a strategy that can be used throughout life. The most common one is the hundreds chart which has all the multiplication facts for 1-10.

- Using number bonds/fact families' templates by the student can enable them to tackle problems that are normally difficult.

- When unsure of the skip counting sequence and/or multiplication facts for a specific number, the student writes down the number to be used, adds TouchPoints, and uses repeated counting to build themselves a skip counting sequence and the multiplication facts for that number. It should be noted that the use of adding TouchPoints to a number, finger tapping, and other learning strategies does not prevent acquisition of math facts or automaticity, strategies such as these are used as a support system and are dropped when no longer needed (Vinson, 2005).

- Use mnemonics for the order of operations, process for long division, and whole part relationships for fractions and ratios.

- Use of a number line to assist with addition, subtraction, rounding, greater than, and less than.

Accurate Mathematical Reasoning

Mathematical reasoning is the process of deciding, using critical, creative, and logical thinking to determine the effective, efficient and logical way to solve a problem (Erdem & Gürbüz, 2015). It is a critical skill for success in math and other STEM fields (Bulat et al., 2017; Espina et al., 2021). Accurate mathematical reasoning refers to the process of using logical and precise methods to arrive at correct solutions to mathematical problems. Key elements include:

- Understanding mathematical concepts: To reason accurately in mathematics, one must have a solid understanding of the underlying concepts and principles. This includes understanding the definitions of mathematical terms and symbols, as well as how to apply these concepts in different contexts (Yoong et al., 2022).

- Using appropriate mathematical methods: Accurate mathematical reasoning also involves using appropriate methods and techniques to solve mathematical problems. This may include using formulas, algorithms, or mathematical models to arrive at a solution. It also means showing the student how to solve the problem and showing incorrect solutions to aid understanding

- Checking for errors: Paying attention to detail and the ability to identify and correct errors in one's work. This may involve double-checking calculations, verifying assumptions, or checking for consistency with previous work.

- Communicating results clearly and concisely: This may involve using appropriate mathematical notation, providing clear explanations of the methods used, and presenting results in a way that is understandable to others.

Instructional interventions that have shown success in improving mathematical reasoning are:

- Logic games, such as Sudoku and crossword puzzles, can help children develop their logical reasoning skills by improving their ability to analyze relationships and make connections between different pieces of information.

- Explicit, systematic instruction of problem-solving strategies, such as breaking a problem down into smaller parts and looking for patterns, can help children develop their logical reasoning skills by teaching them how to analyze complex information and make connections between different pieces of information.

- Math talks involve presenting children with open-ended questions and encouraging them to explore different solutions. This can help children develop their logical reasoning skills by encouraging them to think critically and analyze relationships between different pieces of information.

- Number sense activities, such as counting, skip counting, and number patterns, can help children develop their logical reasoning skills by improving their ability to analyze relationships and make connections between different pieces of information.

- Mathematical reasoning tasks involve presenting children with complex problems and asking them to solve them using logical reasoning. These tasks can help children develop their logical reasoning skills by teaching them how to analyze complex information and make connections between different pieces of information.

- Tangrams are a set of seven shapes that can be combined to create a variety of different designs. Using tangrams can help children develop their spatial skills by allowing them to manipulate shapes and visualize how they fit together.

- 3D visualization exercises involve presenting children with 2D drawings and asking them to imagine what the object would look like in 3D. These exercises can help children develop their spatial skills by improving their ability to visualize objects in space.

- Block building involves using blocks to create structures and designs, such as Legos or Kinex. This activity can help children develop their spatial skills by allowing them to manipulate objects in three dimensions and visualize how different shapes fit together.

- Spatial puzzles, such as jigsaw puzzles and maze puzzles, can help children develop their spatial skills by improving their ability to visualize and manipulate objects in space.

- Overall, these interventions can help children develop their logical reasoning skills and improve their math proficiency. It is important to note that these interventions should be tailored to the individual needs and learning styles of each child, as each child is unique.

Effective learner strategies for students struggling to demonstrate mastery include:

- Using self-checking of the math problems through inverse operations, such as addition and subtraction checks (5-2=3 and 3+2=5)
- Creating a story map of the problem and talking aloud through the steps
- Looking at a completed Frayer-chart and trying to identify any errors; error analysis
- Using highlighters and colored pens to separate unnecessary information from what is needed to solve the problem

Chapter Seventeen

INSTRUCTIONAL STRATEGY INTERVENTIONS

While improving the student's math skills in the four targeted domains, there are instructional strategies that research supports as being especially effective. These are not all of them but rather those that appear to work time after time in most instances, so they are a great starting point. These evidence-based actions or interventions that the educator can take include some of the following.

Systematic explicit instruction or just explicit instruction has been found to be the most effective approach for students with dyscalculia as well as other populations (Chodura, et al., 2015; Dennis et al., 2016; Doabler et al., 2019; ERIC, 2022; Mononen, 2014; Powell et al., 2013) especially when focused on basic arithmetic competencies. It is also known as I do; we do, you do or model, lead, test. Describing it as intentional is a good way to describe it, it is not 'waiting for a teachable moment.' Of course, if a teachable moment happens, grab it, take advantage of it but just waiting, there is no time for that. Teachable moments are a time-honored and effective technique but hard to plan for, so you need to watch for them. Explicit instruction is planned, a goal is set, it does not need to be difficult, it might be as simple as "We need to go over the four facts and I can do it by flashcards and dividing the cookies in 4 equal groups or the collection of toys into 4 different groups". Another way to describe it is as purposefully designed instructional interactions focused on the foundations of math (Hughes et al., 2017). Explicit instruction includes intentional, overt teacher modeling, structured opportunities for students to practice the skills, concepts or procedures that have been modeled by the teacher and then immediate feedback on the practice that the student has completed. We know from general education or Tier 1 classes that this is particularly effective for K-2 classrooms and for those who are at-risk or have math disabilities impacting mathematics concepts and skills (Clements, et al., 2013; Doabler et al., 2019; ERIC, 2022).

There is a time and place for the use of exploration and playing around to find solutions and that can be found in the math curricula used in the general education classrooms your child attends. But when a child has yet to develop the neural pathways that connect the various regions of the brain that process math it is time to use explicit instruction. The use of explicit instruction is the most effective strategy that research and evidence have found for use in cases where new material needs to be learned, and teachers are taught to use it in their methods classes. It is not the only strategy but one that parents can use naturally and with ease.

One strong caution here, please do not think you cannot do this. If you are a parent, you may not have been trained in explicit instruction, but it is really common sense, and you have probably done this naturally. If it is not perfect, it does not matter. The key is you are showing your child you are there for them.

Explicit instruction for struggling students and those with dyscalculia contains instructional elements that include:

- Focus on the key components of a child's needs which are usually the vocabulary, procedural skills, and concepts of math,

- If you can find a way to do this in a real-life manner, do it. If this is about fractions, do real fractions, such as dividing cookies or pizza across multiple friends. It is even better when you get to fractions such as dividing 2 cookies across 5 individuals.

- Sequencing the math that is taught so that the interventions match the math progression. If the student has yet to master addition, teaching multiplication at the same time as addition is not advisable as it develops after addition. So, teaching the easier skill first, which is also the foundation of the later skill is recommended, especially in the earlier grades. After around 5th grade, the use of calculators allows the child to not know the facts, but the concepts will need to be reviewed as they will be the building blocks of later concepts. Not being able to immediately answer 7 x 8 does not prevent a child from solving an algebra problem such as 7(8x) + 3x =121.

- Making sure the child has mastered the prerequisite skills which usually means not starting where a child is struggling but going backward until you can start with the skill in the progression where the student was 100% successful and move forward from there. How do you find the prerequisite skill? Most curricula have a scope and sequence that shows what is taught first. Pull those little pretests and ask your child a few of the sample questions – if they get them all correct, try a few from the next skill in the sequence. If that is where they start to make mistakes, go back to the previous skill, and move forward from there. It is important when a child is struggling, has feelings of anxiety and a "can't do" attitude, that they start from a point of success. If they can do that skill, they can do the next one which is only a little bit harder.

- Teach in small steps. Most commercial or free materials that are available from reputable sources teach in small steps so that the child can master skills frequently, it is much more rewarding to see success every day than only once a week.

- Describe what you are working on for your child. This does not have to be a formal goal statement. It can be as simple as 'let's work on subtraction today, you did great the other day so let's try some more.'

- Do a quick review of what they did before then introduce what you want to tackle today.

- Look at and try to use the vocabulary that engages in the math you are doing for the intervention and do your best to use it – take away is not a substitute for subtraction. You want to try and use the same language as they will hear in the classroom.

- Model for your child, especially if it is new information. You need to show your child through modeling, showing them how to work the problem themselves. When you do this, do it as an example and let them hear you talk about solving the problem, show all the ways you might think about solving it if appropriate, then partially do a problem and finish it with them. If that was successful and you think they, have it, then have them do it by themselves with you watching. If they are not ready to do it independently, model a different problem, work one together and then let them practice.

- Modeling means lots of examples and especially non-examples. Examples of what not to do are just as important as what something is. Imagine talking about squares and describing them as having 4 sides and each angle is 90 degrees. If you do not share what a rectangle is they could easily make a mistake.

- Give your child a lot of opportunities to respond. This can be orally, written or shown to you. With math, this means build it for me with manipulatives, show me with manipulatives how to do it, draw how you would solve this using circle, hash marks, stick figures. And if they can work with the numerals either with the algorithm or if they have a different way of working it. Have them talk you through it as they do it.

- Provide immediate verbal feedback. This should be as soon as possible and should not use the word wrong. Rather say let us try that again, show me how you got that answer.

- Do not do these activities for long periods of time. Interventions can be just a few minutes. If you think about it, you do these all the time with your child and for more than just math. Try to stay under 20 minutes unless the student is totally engaged and now you have found a magic teachable moment! Take advantage of it but stop when the child starts to disengage.

- Take a deep breath and relax. Interventions as a parent should not be anxiety causing for you or for your child. In the case of these, the school is also providing interventions, and you are the extra help that makes a huge difference but always remember-something is better than nothing. So, every little bit you do tell your child you are there for the hard work and they can succeed. That is a positive step forward.

- Explicit multisensory instruction should act to increase motivation and boost conceptual knowledge gain (APA, 2022; Mahmud, 2020). This includes dance which involves the whole body, as well as choral response, which is not only engaging and provides for engaging practice but also is known to help the brain retrieve facts. It also includes the use of presenting visual, auditory, and tactile sensations concurrently as is used in the TouchMath materials (Abdou, 2020; Taneja, 2019; Urton et al., 2022; Vinson, 2004; Waters & Boon, 2011; Wisniewski, 2002).

- Immediate verbal feedback helps with math facts (Fuchs, et al., 2012)

- You are probably wondering about Computer Assisted Instruction (CAI). Used properly it can adapt to the child's learning speed as students with dyscalculia typically need more structure and time. CAI also provides immediate feedback and the gamification is an important engagement tool for many students (Chodura, et al., 2015; Kohn et al., 2020; Mahmud, 2020; Mononen, 2014) In addition, because it is 1:1, there is the ability for it to be non-competitive and reduce fear of failure as the child is competing against themselves, thus reducing math anxiety. This also includes the use of graphing software as well as gaming software which allows for unlimited practice as well as motivation due to its high engagement factors.

- Retrieval practice – Retention of knowledge and skills is higher when practice is spaced across multiple sessions instead of there being only one practice session as in cramming for an exam. It is commonly accepted that last-minute study can provide enough retention to pass a test but if asked to recall or use the skills days later much of the information has been lost. Current research is supporting that an individual may be able to learn something quickly but without repeated sessions to retrieve and use the material it is quickly forgotten, and the strategy of spacing practice and retrieval sessions is especially important for those who already have difficulty retrieving math facts (Walsh, et al., 2023).

- Extensive rehearsals – especially for fluency and automaticity. All students benefit from multiple opportunities to do or practice a new concept or skill until mastery, including those with dyscalculia. Initial practice sessions should be closely supervised to ensure that the sessions result in reinforcement for accuracy the first time around and that errorless learning is implemented (Bulat et al., 2017; Espina et al., 2021).

- Timed drills where the student is competing against their own time only and results are immediate and reinforced appropriately are effective for increasing processing speed for retrieving math facts and reinforcing as the student only competes against their own time (Mahmud, 2020).

- Errorless learning and wait time provide time to properly process new information or a problem (Mahmud, 2020), minimizing learning challenges. Another intervention, called mathematical discourse, is the communication styles in the classroom, utilizing strong math vocabulary and teacher-student, student-teacher, and student-student discussions about mathematics that lead to a conceptual understanding of the curriculum being practiced is effective for mathematical problem solving and reasoning.

- Cumulative reviews are a strong best-teaching practice. A strong cumulative review ties in different areas of the curriculum and helps students create conceptual frames around the structure of the dependencies of mathematics.

- Concrete-Representational-Abstract (Bouck et al., 2018; Jacobsen, 2020; Mahmud, 2020; Mononen, 2014). Just as we do not want to teach concepts separately from the procedures, the use of manipulatives, providing concrete and visual aids for problem-solving and practice enables students to connect abstract symbols and numerals to their numerical values and experience the concept and the procedures that connected to it. The use of color is encouraged as it adds an extra element of being engaging to any age student. The use of dice, Touch Numerals, etc. allows student to directly connect the concrete and the abstract. Using cereal, beads, tiles, and chips also enables a tactile technique for engaging the students and explore the math concept or demonstrating their understanding of it (Cihak & Faust, 2008; Yikmis, 2016; Ellingsen & Clinton, 2017; Fletcher, 2010; Kot et al., 2018; Taneja, 2019).

The list of effective instructional strategies that are research backed has increased over time (Hattie, 2023) and we have highlighted the ones we have found most effective and easy to implement by teachers and parents. Other strategies include peer-assisted learning (Mononen, 2014) teaching examples and non-examples, providing sample problems, explicit vocabulary instruction to make it easier to recognize words tied to a problem, the Universal Design for Learning (UDL), on-going progress monitoring, such as Curriculum Based Measurements.

Chapter Eighteen

LEARNER STRATEGY INTERVENTIONS

When a student has dyscalculia there will usually be areas of mathematics that they will struggle with permanently. These are the actions that the individual with dyscalculia uses to do math as well as the individual who does not have dyscalculia. Think about the adults you read about in Chapter 3 and the actions or strategies they use day to day. We cannot and must not assume that the child will pick these up on their own, we must deliberately teach these strategies and have the student realize these are their glasses if they near sighted. Please note I also covered a number of these a few pages back so you may want to look back at the interventions for the math domains.

The use of a smartphone is one of the most frequently used but if that is not available and the student is in the process of mastering the skill there are alternative ways of approaching the problems that will allow them to solve it in a less efficient but still successful manner.

Learning strategies can be taught as temporary or permanent support as they learn to master various math concepts and skills. These include:

- Taking time to review work, deliberately teaching students to slow down, and read a problem twice, asking themselves if the answer is logical can help them avoid errors.

- Using templates and graphic organizers such as Frayer charts and

BiDiWi templates helps students.

- Having students solve a problem and then use the inverse operation to check their work decreases mistakes but also gives the students additional methods to solve problems.
- Writing down the steps before attempting to solve a problem (Kroesbergen et al., 2022)
- Having students take large tasks and break them down into smaller tasks and then complete them in order (Arizmendi et al., 2021; Kroesbergen et al., 2022).

Instructing students in schemas; the frameworks, outlines, and plans that are used to solve problems is also an effective intervention (Driver & Powell, 2017). Students can use the schemas to organize information from a word problem in ways that represent the underlying structure of the problem.

Explicit sharing of the scaffolds that teachers have put in place for a struggling student can enable the student to have a better understanding of how and why they are struggling and the supports they can use to help themselves as they work toward mastery. An example would be the teacher removing distractions from a room to aid a student who is easily distracted. The student in turn could deliberately choose settings to complete their math work that they recognize as not being distracting.

In Summary

I hope that you have found these suggestions to be not understandable but also easy to implement in the classroom and at home. And please check out the resources in the resources section for additional ideas that we have identified from other sources and educators.

Chapter Nineteen

LET'S DO 'MATH'!!

"Let's do math." A very familiar phrase in a lot of classrooms and at homework time for families. This chapter gives you activities to do with your child that will help develop the areas of common struggle.

Support your child emotionally as math is hard work just as learning to read can be if you have dyslexia. Be positive and try these suggestions.

- Activities organized around the 4 primary areas of struggle for students with dyscalculia: number sense, development of math facts, the basic calculations, mathematical reasoning and problem solving.

- Activities for some areas that are common areas of struggle - time, money, and directions.

- Activities for areas that often are not thought of as math but are necessary to successfully master math concepts and doing math. These include visualization, visual and spatial working memory, and math vocabulary.

- Activities for the areas that enable the math portions of the brain to perform their work. These include working memory, attention, processing speed, nonverbal reasoning and phonological processing,

Getting Started – Things to do to support your child as they work hard to learn math

There are actions you should take as the adult when you do these activities with the child.

- Encourage your child to do positive self- talk. Have your child tell themselves- "I can do this," "I am good at math," "I can learn from my mistakes," and "It's ok to need more time, I got this!" Positive self-talk can help boost confidence and reduce anxiety. As the important adult in your child's life, you need to say these statements to them and to yourself. YOU GOT THIS!!!

- Teach your child to take a deep breath. They can also close their eyes and see themselves doing the problem. Persistence pays off.

- Keep directions clear and short. Expect to remind them what the next step is. Break tasks into small, manageable chunks and keep the time you are working with them short unless they are eager to continue, 10-20 minutes is a great rule of thumb. Go do some physical activity and I bet you can even find a way to do some simple math at the time, or directions or math vocabulary. Completing the first 3 steps is always easier than doing a set of 10 steps. Make your sentences shorter or use a blank piece of paper to reveal information slowly.

- Provide enough time to work a problem or process information. This is especially important for new information or a new skill or concept. Providing enough time can mean watching your student and counting to 10 or asking if they need help. Observe how long they typically need and provide that time for them.

- Practice verbal rehearsal out loud. Talking in a soft voice that your outer ear hears, not the voice in your mind, is important to learning as it is a completely different set of neural pathways. This helps store math facts, number names, etc. in working memory.

Have the student talk softly but definitely aloud when doing independent work. This may mean finding them a separate place to work.

- Establish a consistent routine for doing work. Scheduled time for doing math, play, and other focused activities is easier as the child can prepare themselves for it and not be in constant anticipation that a parent or teacher will suddenly say Let's do Math. For most of us that would be akin to a pop quiz. For this child, it is probably a high anxiety event.

- Point out all the times they are doing math and doing it well. Deciding which to do first, eat the chicken nuggets or drink the milk is doing math as you are using the math vocabulary words of first and second. Asking them to tell you a story has them putting events in order. What happens first, second. What happens next. All of that is practicing sequencing, an important math skill.

- Have fun with counting. It doesn't hurt to add in the math fact as you do this but don't ask them to guess what 4 + 2 is. That is for later after they have started developing the concept that 4 is always 4. In addition, having them guess is not something they should be doing. Based on what we know of neuroplasticity, guessing and getting wrong answer means the child will have to work harder to 'forget' the wrong answer as they learn the fact.

- Use manipulatives for these first activities as math understanding starts with the concrete world and understanding that the abstract symbols – 3, 13, and words – three written or spoken; are in fact referring to real 3-dimensional objects. And the manipulatives actually let a child take 7 objects and put them into a group of 6 and 1, 3 and 4, or 7 and nothing, and understand they are all the same number for those 7 objects. And keep using manipulatives and drawings to have your child show you how they would solve a problem.

- Don't tell them they are wrong. If they make a mistake, simply say the correct answer. For example- Count the green beans for me. If the child responds 1,2,3,5. Simply say – 1,2,3,4. There are 4 green beans. Try again for me.

- You, the teacher, should verbalize, say the steps yourself. You are talking out loud and have your child do so also. Saying 4 plus 2 = is 1,2,3,4,5,6; 4+2=6 helps you and your child do this as a conscious action and hear their thinking. If this is a classroom, then other children also get to hear the procedures and may learn new strategies. Remember choral practice, let's say it together and you can fade your voice as they become more accurate in their answers.

- Listen to your student or child- by hearing what they say you can quickly judge if they are understanding the procedure or if you need to adjust the difficulty level of the problem or the speed expected for an answer. Ask them to explain what they see or think they should do as they are solving the problem.

The four primary areas of struggle and suggestions on how to help with the math.

Number sense

If we revisit our description of number sense we know that it is the ability to understand and work with numbers, to think flexibly with them rather than just memorize a single algorithm. Number sense is about playing and experimenting with numbers so that you know what you can and cannot do with them and have multiple ways to solve a problem. An example is that 5 blocks are always 5 blocks regardless of whether one stack has 2 in it and 3 in the other or there are 4 in one stack and 1 in the other. The development of number sense begins before age 3, builds over the course of a child's primary and elementary school years and is the foundation for later math development in middle and high school. If a child starts behind and there is no early intervention, it becomes harder and harder to keep up with peers as the math continues to build (Clements & Sarama, 2021). It can be done but why put a child through the pain and frustration that you read about in the case studies of adults who found out after high school that they had dyscalculia.

Activities to help develop number sense for the younger child

To help children with dyscalculia develop number sense there are a number of activities that can easily be done by the parent at home or by the teacher during the school day. This is only a sample as there are any number of activities available to help your child that can be purchased or available free online.

Learning to Recognize Numbers and Count is the First Step in Number Sense

Humans have used their fingers to count, and our base ten system is founded on the use of ten fingers. Teaching children to count on their fingers is one of the first skills we teach and that each finger can represent a numerical word or symbol as well as a spoken word.

If a child has not yet learned how to count and be fluent with their math facts and early operations (2+2+3), finger counting is an intermediary step until mastery that allows the student to successfully solve a problem and alleviates the demands on working memory and the triggers for anxiety. Fingers also act as visual, motor and tactile cues. So, learning to count and use our fingers is one of the most crucial early skills as the quality of the finger representation or usage has been shown to be a better predictor of math skills than standard developmental tests. Finger counting sets the stage for learning addition and the other operations, helps the student to develop the understanding of 1:1 correspondence and recognition that numbers are made up of subgroups. Children with dyscalculia and other mathematical learning difficulties rely more on their fingers for a longer time period than their peers to do math facts and operations. (Berteletti, 2015; Geary, 2013). Many of us actually still use our fingers when asked to recall certain math facts, with 6 x 7 and

7 x 8 being two common ones. Adults who are good at math describe themselves as having to recall the 7 x 8 fact as 7 x7 is 49 and then begin counting up, but they recall that the answer is 56 before I finish my discrete finger tapping.

Think of using the fingers as a backup method for figuring out a math fact or problem when you can't immediately remember the math fact and your anxiety level begins to kick in. And recall of math facts and anxiety are common for those with dyscalculia. And not to worry, this technique will naturally fade away as the child becomes more fluent with their facts because it is essentially an ineffective technique for backup only

The First Step- Subitizing

Subitizing is the ability to recognize the number of objects without having to count them. If you look at the figures below you do not need to count the balls or the birds to state without hesitation that there are 5 balls and 3 birds. A child with dyscalculia may not be able to do that, they may need to count each object before being able to state that there are 5 balls and 3 birds.

Find opportunities to have your child tell you how many without having to count. When the number is 5 or less it is usually recognizable. Placing 3 spoons on the table should generate the answer of 3 without having to count. If your child struggles start with 1 or 2 items and asks how many, then gradually move to 3, 4 and 5. Most of us need to count the items if they exceed 4.

And pull-out groups of items that have a very recognizable pattern and do not need to be counted. This would be dice and playing cards which can usually be recognized without counting to 12.

Counting

Things you can do to help your child with counting are:

- Start with their fingers as our base 10 system is built upon the early use of our fingers. Have your child hold up one finger and match it to 1 object. Do the same for 2 and gradually build until you get to 10.

- Begin by saying the number, writing it, having the student trace the number in the air.

- Put 5 blocks, plastic ducks, straws, whatever you have on hand on the table. You count them out loud, then count them with the child out loud, let them count them out loud, then have the child draw the same number of items on a piece of paper or a tablet and then write the correct numeral, 5. Do this with other numbers. If they make a mistake do not say 'wrong.' Simply say the correct answer statement. "1,2,3,4,5 ducks. There are 5 ducks."

- Play games such as "I spy with my little eye" 3 towels in the bathroom or 2 toothbrushes. At this point don't add other identifiers such as color.

- Sing counting songs such as "1, 2 Buckle my shoe" and the "Ants Go Marching One by One" and act them out with your child.

- Find books to read and to color at the library, local stores, or online. Coloring books are also great as they involve fine motor skills which translate to the writing of numerals.

- Point out where there are that many of the number- open the refrigerator and say, "I see 3 carrots," point out other numbers in the real world at the grocery store or gas station.

- Present objects in a cluster such as 4 green beans and 2 green beans and count all of the green beans. 1,2,3,4 green beans, 5, 6 green beans. Can you count with me as we touch them, 1,2,3,4,5,6 green beans. See if the child can continue and do it again on their own.

- Use dominoes to perform the same problems and then move to dot cards .

- If you have playing cards pull a 4 and 2 of same suite and count the hearts or the clubs. Once again start with counting all of them, have your child do it with you and then have them do it alone. If they are getting it, you could add the final sentence of 4 + 2 = 6.

If you are making dot cards, try to make them have a recognizable pattern in order to reinforce the development of subitization and a connection to the math facts. Examples would be:

- Card A is a classic 5 dot pattern that can be instantly recognized whereas Cards B and C demonstrate number bonds- that 5 can also be a 4 and 1 or a 3 and 2. This same exercise can be done with numbers 6-10 on one card but larger than 10 and it is probably easier to use multiple cards.

- Look for card games that reinforce counting or the game of dominoes. There are games for students as young as 3 and as old as adults. Key thing is to make it fun and involve other children and adults. Use the computer games for short periods of time as the child needs to see and hear others using numbers and talking through strategies.

- Look for memory games where the cards or objects are hidden, and everyone needs to match the pairs from memory

- And don't forget games such as Chutes and Ladders, dominoes, etc.

- Put objects or actions in order of their use so that children can describe what is first, second, third.

- Have them count objects such as 3 blocks, 2 balls, 4 sticks, or cardinality and then ask them to put them away in a specific order, "Put the balls away first, then put the sticks away second, and the blocks will be third.' This helps build and reinforce that there are words for the order of things, ordinality.

As time passes and the child becomes more accurate with their counting then speed can be added by showing the child items to count for a few seconds and seeing if they can count them accurately.

A Key Part of Number Sense-Magnitude and Comparison

Being able to determine magnitude is another important math skill that is easily addressed and can be a lot of fun. Numerical magnitude processing as it is called in the academic world is the ability to determine which group of items has more items in it or has the greater quantity. This can be around size, density, area as well as quantity.

It is also the ability to look at a list of numbers and determine if the order of the numbers is getting larger or smaller.

You can also do this with number lines so they improve their cardinality and ordinality skills. Using a large ribbon that you have at home you can do the same activity. The child can make their steps be the 10 numbers and start with 0, 1, 2, 3, ... until 10. By using their own feet which are a constant size it helps them visualize that the numerals are equidistant from each other.

The next activity is to have them draw the 0 to 10 number line. Have them put the numbers on the line and have them put the 0, 5 and 10 first. Then have them count to 10 and fill in the rest of the numbers. You are looking to see if they are beginning to understand the relationship of the numbers to each other.

Using a 5 or 10 frame is also helpful. You can draw a 5 or 10 frame on a piece of paper and have your child number them, this keeps the numbers equidistant. This is a great activity to do with manipulatives by having them place an object in each frame and name it, they can also write the number in the frame. For more fun use an old egg carton and cut off the last 2 egg holders. Using manipulatives really helps the child understand that zero is nothing, there is nothing there.

Start adding some magnitude questions by asking them which number is larger 3 or 7. Have your child place the numbers and describe how the further right you go on the number line the greater the number. It is also good to do this with a vertical number line and compare it to an elevator or stairs.

Working on Estimation, Another Key Component of Number Sense

Estimation is essential to doing math. Imagine not being able to estimate whether an answer is close to accurate- the difference between 10 x 12 being 120 or 1012. Estimation is that feeling one gets when you are pretty sure the answer is not right, but it is an important part of number sense.

There are some fun ways to work on estimation and magnitude at the same time and especially with manipulatives that are commonly found in the home or classroom.

Select an item that you have a fair number of in your home or classroom but for the first exercises try to use items that are exactly alike so that the child is not distracted by sizes or colors of objects. This could be spoons, paper clips, rubber bands, etc. Make a big pile and a very small pile and do an I do, we do, you do exercise.

I do- It is important for your child to hear your thinking and how you work the problem so speak out loud and let them hear your chain of thought. "Wow, which group of spoons has more – I think this one does because it looks like there are more. How do I know? Well, I can match them." Proceed to match 1 spoon to 1 spoon between the groups until the smaller group no longer exists and talk about how there are more left. Continue the process by starting over again and having less of a difference between the two groups and repeat the 1:1 matching process. The last step is to have the groups be equivalent and do the matching until both piles are matched and say "Now the groups are equal"

We do- change items or use the same ones, only this time do the matching together. You take one spoon, and your child takes the other and start the sentence of "Look this group is larger because there are more ..."

The last step is to have the child look at the two groups and tell you which one they estimate to be larger and then actually do the 1:1 matching to prove that they were correct.

This is an activity that lends itself well to spreading over several days which increases the ability to retain the skill and knowledge long term. As your child becomes more comfortable you can phase out the matching unless the 2 groups are similar in size.

As your child gets comfortable comparing the groups of the same objects start using mixed objects such as spoons and forks, spoons and balls, different colored items. The goal being for your child to understand and consistently be able to tell you when one group is larger than the other by simply looking or starting to do the 1:1 matching and to explain to you why they think one group is larger.

Find normal daily activities where you can look at a group of cars at a stop light and ask "wow, which stop light has more cars at it" or during a walk "which bush has more flowers or birds?" etc.

Another activity you can do is select between 5 and 10 objects that are all the same and pile them up so the child cannot subitize or recognize the number without counting (usually 4 or less can be easily recognized). Using the word estimate, ask your child to estimate how many objects there are. Then have them count and see how close they are to the actual number. Wild estimates should be responded to with "count out 3 of the objects, now compare them to what is left, do you want to change your estimate?"

Activities to Help Develop Math Facts for the Younger Child

Learning the math facts for addition, subtraction, multiplication, and division is a critical part of building the math foundations needed to master more advanced mathematics. It is of course possible to never learn the math facts but that would probably leave a student with having to rely on their fingers or basic counting skills to solve all equations. Doable, but not particularly effective or efficient. We know from the research that learning the facts is particularly effective for the child with dyscalculia (Chodura et al., 2015; Dennis et al., 2016). The adults I interviewed are still working on their math facts and do not find it to be a waste of time as knowing, any or all of them, makes it easier to do the math computation side of problem solving. Relying on a calculator does not work as many individuals with dyscalculia can enter the digits incorrectly due to working memory issues or not recognizing that the answer is wrong because of estimation problems.

The ways that you can help your child as they work on the memorization of facts often can use the same materials used for the development of number sense. TouchCards, flashcards, computer games, dominoes, manipulatives, etc. are all useful tools to have.

Use flashcards to start with the simplest facts. The 1, 2, 5 and 10 tables are the easiest to learn. When teaching math facts, make sure that the first few times your child hears the fact they hear the full fact and the correct answer. The suggested dialog goes something like this:

Adult script

- Let's learn our addition facts for the number 2.
- 1 + 1 = 2
- Say it with me, 1+1=2. If your child appears to be struggling some. Repeat the math fact again.
- Now say it with me, 1+1=2
- 1+1=? This final time, you say the fact but do not provide the answer.
- If they answered correctly, feel free to say "correct, 1+1=2."
- If they did not answer correctly, repeat the fact 1+1=2 but you don't need to tell them it was wrong as your stating the fact again with a number that is different likely makes it obvious they had the incorrect answer.
- Use those manipulatives and drawings. Typically, math facts are thought of as flashcards, but practice is the key to memorization in this case but understanding that a math fact is a shortened version of the building out of the manipulatives or drawings of that many items. Have them practice the fact with the BiDiWi.

For addition, the suggested dialog might sound like this:

- Let's do the addition math facts for 2. 2+1=3
- Here are the blocks. 2 blocks + 1 block + 3 blocks. 1,2, 3
- Then have the child count with you.
- Then have them count alone.
- Then have them draw the problem using any picture they want to represent 2 and 1.
- Then have them write the math fact.
- These same facts for all four operations can be found in a wide variety of computer games which most children find engaging. The key may be to limit the time on them.
- Board games can also be used, but as the adult you may need to call out the fact. Rolling a 2 and 1, you can share that that is a math fact they just learned; 2+1=3.
- Have your child visualize the fact in their head. Can you see 2 dogs and now another dog joins them. Count the dogs in the picture- 1,2,3 - 2+1+3.
- Have your child build their own flashcards.
- Increase the speed of your expected response as accuracy comes before speed. Use quick "Beat your last time" games but don't have the child compete against others. Keep a big chart showing their progress as the line moving ever upwards demonstrating progress is a major motivator.
- Buy or build your own 100's chart as it helps children see the pattern of the math facts, use it to practice skip counting forward and backward as it will pay off for learning multiplication and

division facts.

Activities to Help Develop Math Operations

Math operations, addition, subtraction, multiplication, and division, come naturally to children even before they have the numbers to work them. Think division, which most of us would consider to be the most difficult of the operations. Try breaking a cookie in 2, with one piece obviously larger than the other and offer the smaller piece to a 2-year-old. You really think they don't notice that it was not 1 cookie divided into 2 equal pieces. Think again!

Of course, understanding and being able to do the operations are critical to math achievement and highly predictive of future math mastery (Fyfe, 2019; Kim et al., 2022). As with the math facts, being able to solve the problem correctly is more important than the way it is solved or the speed with which it is solved. Eventually students acquire the most effective and efficient means of solving problems correctly as that is the nature of humans. Remember, accuracy first, then speed.

Learning the operations of addition, subtraction, multiplication, and division are part of the typical elementary school daily routine but there is a lot that can be done at home or by teachers whose students are struggling to master the operations. Some examples are:

Dominoes and other board games are particularly effective for addition and subtraction as the ability to move forward or backward on a board is the heart of addition.

There are computer games also that are directly aimed at mastering the operations and provide the gamification and fun that make this form of practice enjoyable and not seem like practice.

Pull out that BiDiWi and the operations can be performed with manipulatives, directly tied to the drawing of the problem and then the abstract. Think dividing a pile of manipulatives by 3's, 4's, etc. and you have reinforced division.

Use real-world problems, how many glasses will I need if we have a party for 6 people and we need water, juice, etc. and each must be a different glass.

Help think of or google mnemonics for the operations so that the steps are more easily recalled by your child.

Use number lines, TouchPoints on numbers, finger tapping or pencil touchpoints to help solve the operations when your child does not have the math facts easily at hand.

Activities to help develop mathematical reasoning for the younger child

Many of the activities you have been doing naturally with your child while driving, shopping, and walking, and those you have been doing deliberately to provide the extra time and extra experiences with math help in the development of mathematical reasoning. The ability to look at a problem, determine ways to solve it, select the most efficient method and then explain their logic to you is something you have been doing every time you ask them why or how they did it that way.

The sorting games you played taught them putting things in order from smallest to largest, seeing the patterns in a string of beads – red, 2 yellow, blue- are all aspects of mathematical reasoning. Other ways that you can help develop this extremely important skill that is so predictive of later achievement (Green et al., 2017), are:

- Have them use language to talk through what they are doing; make need to share your thinking out loud as you solve a problem. Which group is larger? Your answer could be along the lines of – if I look at it, that one looks larger. But if I put them side by side or one on top of the other, this one is actually larger.

- Moving from simple problems to more complex problems- reasoning develops over long periods of time and comes from solving lots of problems.

- Make problems real life. Use your child's name or that of family members in the problem. You can also make the problem about something that is going on in the classroom or neighborhood.

- Show them that some information is not important. "We went to the store and bought 4 cookies and ate 1 on the way home. But 3 of your friends came over. How will we evenly share the cookies among you?" The store is not important but figuring out how to evenly divide the cookies is essential.

- Logic games, such as the simplest Sudoku and crossword puzzles help children develop their reasoning skills by improving their ability to analyze relationships and make connections between different pieces of information.

- Problem-based learning and open-ended questions enable students to. Not having to solve the problem can be a fun activity. The child may also feel less anxiety if they know they don't need to solve the problem, just talk about it. Or the child can solve it in any manner they choose and there may not be a right answer. Flexibility is important as some students 'need' to know there is an answer and for others that is not the case.

- Jigsaw puzzles, mazes and tangrams help the child develop spatial skills and allow them to manipulate objects in 3 dimensions. It can also be done via computer with virtual manipulatives.

- Legos and K'nex also are enjoyable, easily accessed and develop spatial and reasoning skills.

Math reasoning is an area where having the child develop the practice of self-checking by doing the opposite operation is important, 5+3=8 and 8-3=5. Same for multiplication and division. Or working the same problem using a drawing or manipulatives or doing math a 2nd and third time. The adults interviewed had a common practice of checking all important math problems a 3rd time, not 2 times to make sure it was right. They also would talk through the problem aloud and answer softly as they could spot that it 'didn't seem right' and then get help to verify.

Other Areas of Math That We Don't Think of as Math

Let's talk about how we can also address some of the areas of thinking that are used when we do math and typically any subject as these can help your child with reading, science, and other school subjects and will also be used as adults.

Some Activities for Time and Money and Getting Lost Because Directions Make No Sense

Because time, money, and following directions are 3 of the areas that the adults I talked to said were especially troublesome for navigating day to day life I have included some activities you can do with your student to provide supports.

Time – in today's world it is important to be on time to class, to get homework in on the proper day, or to know how to estimate how long to warm up a meal. These are some of the common day-to-day activities that adults describe as frustrating and in need of backup systems so they can be on time. Looking at an analog clock or knowing what 10 minutes felt like was not something they could successfully do. Activities and supports you can provide your child with include:

- Do practice reading an analog clock as well as digital clocks and phones. The easy thing is to ask what time it is. Set reminders by asking the common support systems such as Alexa or Siri to set a time for 7 a.m. if that is the time the child needs to wake up to do everything to get ready for school.

- Because there are a number of things that need to be done to get ready, multiple timers can be programmed across multiple days using calendar features with their labels. 7 a.m. is wake up time, wash and dress, 7:30 a.m. signals that they need to be downstairs for breakfast and gather homework, 7:50 a.m. is noted as the time to walk to school or bus stop.

- But because our days have things that are not set in stone; breakfast took longer than expected or the dog needed to be fed;

it is important to also work on estimating how long something will take. Play games that have timers, have them see how close they can estimate 30 seconds, 1 minute, 15 minutes. Use words such as early, late, or on time based on their answer.

- When working problems get a clock that they can move the hands on and learn the pattern of numbers.

- Video games that have timing features are also a good way to get the brain to develop a sense of time

Money-even though we can use our phones nowadays and make contactless payments if we cannot understand the monetary system and just how much money was spent this month compared to what one makes for pay then getting into debt is a real problem. Adults with dyscalculia have identified overspending and living beyond their means as a very real problem. Banks and credit card companies do not make accommodations after the fact, they tend to not forgive late payment penalties and interest charges, and the credit bureaus do not have special designations for individuals who have difficulty with the concept of money. What can you do to help your child?

- Do teach them the monetary system

- Have them triple check answers, even when using a calculator.

- Use limited debit cards and have them check how much is left before heading out.

- Work money problems and practice estimating and rounding – I have 30 dollars and need to buy this week's groceries. Do I think I can do that with 30 dollars.

Visualization

Another extra area to develop that helps math development is visualization, the process of forming mental images of the problem as we do math. This includes being able to draw a picture of what a math problem looks like, reading that John has 5 glasses of lemonade and needs to give 2 to Beth and then doing a picture of that. We know from studies in neuroscience that the visual area of the brain is activated when working math problems and that they both play a significant role in math performance over time (Fanari, 2019). We also know that the area of the brain responsible for fine motor skills – the fingers -is activated when doing math especially counting (Barrocas et al., 2020; Berteletti & Booth, 2015)

There are several ways to help your child develop visualization skills including:

- Start with finger play games where the child can see their fingers such as show me how many. The child holds up the number of fingers you model for them. Start with 1 hand and move to two hands and make sure to include 0.

- After the child can do these simple activities ask them to show you the number of fingers for the number of objects you show them

- Then move to them not being able to see their fingers- bunny ears are probably the best position as the child can place their hands on the top of their head and show the number of fingers but must now see their fingers in their mind's eye.

Visual and spatial working memory

Working memory plays a role in that the student needs to hold the information long enough in memory to accurately draw it. Even the short period of time from looking at the arrangement of 10 objects arranged in groups of 7 and 3 and then drawing it as 7 circles or lines and 3 circles or lines is a critical step in understanding that 3 dimensional objects can be represented as pictures and then symbols or words. This ability to represent the 3-dimensional object as a picture is critical to developing the ability to work in the abstract, and without it students will struggle with the more complex concepts and skills. This move from representing objects as pictures and then symbols is called the concrete pictorial abstract framework and is credited to Jerome Bruner (1996).

CRA or concrete-representational-abstract is a term also encountered and the two are very similar with CRA being the term used in this book. You will also find this process of students learning how to work with abstract concepts by moving from the use of concrete activities to representations of those concrete activities with visual or drawn activities before using the abstract form. It is generally accepted that this process is how we develop math concepts and procedural knowledge. What is important is that child can see how the concept or procedure works when they build it or manipulate objects including their fingers and then draw a representation or picture of what they did. Moving to the use of numbers or words allows them work problems that are larger and more abstract.

There are two forms of working memory that are predictive of math skills, and they are visual working memory which allows the recall of the memory of shapes or numbers that were seen while spatial working memory has to do with their location and sequences, was it 5-3 or 3-2 when working a problem or transcribing it.

There are a couple of fun ways you help your child develop visual spatial skills that are easy:

- Spatial memory games require the child to remember the sequence of movements. Doing fun exercises- "can you do what I just did" games help with that.

- Games that involve using building block patterns and then having the child build the same pattern but without the original in front of them help improve spatial and visual working memory.

Metacognition is being aware of how we are thinking about doing something. It is knowing that we are thinking. It is knowing that I can solve the math problem of 110-99 by doing a full-blown traditional subtraction problem or I can do it more quickly in my head because I understand that numbers are flexible. I know that 100 - 99 is 1 and then add the 1 to 10 and the answer is 11. Much easier and faster than doing:

$$\begin{array}{r} 110 \\ -99 \\ \hline 11 \end{array}$$

Children with dyscalculia may want to rely on the algorithm even though it takes longer because it worked in the past but having multiple ways to solve a problem is helpful when anxiety hits while taking a test and remembering one's math facts just isn't happening.

Building Metacognition

Some easy ways to help your child be better at metacognition are to:

- Ask your child questions about what they are thinking is the most efficient way to solve the problem

- Have your child think out loud what they are seeing and thinking as they look at the problem, have them plan what to do before they start to work the problem

- Example:
    - What do you think when you see this problem?
    - Have you worked any problems like this before? How did you solve those problems successfully? Can you find that problem again? How did you work it? (Usually this is when the child is working a set of problems or when they have their textbook)
    - What is it asking you to do?
    - Who can I ask for help or where can I go for help? (You are hoping to hear that they have a friend they can ask, a Q&A resource on the district or school's website, a homework hotline, etc.)
    - How might I use manipulatives to solve this problem? Can I sketch it or draw it? Think of the phrase- Build it, Draw it, Write It (use the BiDiWi)
    - What is the first step you want to take? And then? How will you know if the answer is correct?"
- Have them work on the problem and monitor themselves as they do it, have them trust themselves and assure them that stopping and trying another method if they are unsure is what good mathematicians do all the time. The more ways they try the more options they will have when working to take a test
    - Is this working?
    - Am I thinking I am not going to get it correct?
    - Do I need to try another strategy?
    - Do you need to build the problem or draw it out, is there another algorithm to use?

- If I do another problem like this, will I want to use the same strategy because it was the easiest and quickest?
- Did I struggle anywhere, did I get confused? Why?

If you have the opportunity as a parent or teacher to work the problem with the student or to have them work it in small groups so they can compare how each of them solved it, they can learn other ways to solve the problem or have the technique they used validated. Because reflecting and writing are excellent ways to create long term memory, it is important that there be a debrief at the end of the session when you are deliberately working on acquiring new strategies, whether it is building metacognition skills or working a new type of math problem or just cementing prior skills. Simple math notebooks that allow students to write down a few key ideas after a discussion help them remember their successes and what they learned.

Executive Functioning- what is it, what role does it play in dyscalculia and what do I do about it as a parent

As you saw in earlier chapters, executive functioning and the other domain general areas are just as important as the math areas but fortunately research supports that they develop as you do the math (Clements & Sarama, 2021) but they are definitely worth reinforcing as they are just as critical to the other academic areas and life success. They are especially important for the child with dyscalculia as it is an identified area of struggle so more time and exposure to activities that develop executive functioning is important. So, what can you do to support the development of these:

- Help your child pay attention to details, especially those that are math related. When in a store, have fun identifying shapes, counting how many corners you turn, see and point out the numbers. It seems like you are doing a math activity, but you are reinforcing paying attention to details in this activity.
- Those math facts actually help working memory, if you know the

math fact and don't need to think about it, it will not impact working memory as you do more complex problems. Even the adults I interviewed who have dyscalculia have not given up learning their math facts. It makes doing other aspects of math faster, easier and more accurate. Look at the activities suggested in this chapter for working on math facts and do more than just flashcards.

- Build up access to long term memory as we know accessing facts is a struggle. Practice with various problems, flashcards, activities, graphic organizers that enable the child to see a concept in connected ways.

In summary

There is so much that can be done to support what is going on in the classroom before a child is diagnosed with dyscalculia, or even when they do have an IEP or 504. The key is to work with the school, the therapist, whoever is supporting you. Key things you can do are:

- Speak positively of what the student can do and give them time.

- Practice makes perfect but make it as real life as possible and don't make it a punishment. Make it as fun as possible and keep it to no more than 10-20 minutes.

- Point out how math is used in day-to-day life and reinforce that they do math, and it is hard work.

Acknowledgements

While researching and writing this book, I realized how much I owe to fellow educators, researchers, parents and individuals in the world of exceptional education. This book is dedicated to them as they have helped me to understand the world of this ridiculously unknown and underdiagnosed disability, dyscalculia. They pushed me to share what I have learned and provide simple, practical advice. And all of them have been so patient as I talked their ears off about the latest research and the various paraphrasing I could use to translate into laymen's terms. I can't thank them enough as they listened, responded, laughed, gave me puzzled looks and made it possible to write a book that I hope helps my colleagues better understand and provide the proven interventions that can enable those with dyscalculia to master math as well as their peers.

I would like to acknowledge my colleagues at TouchMath, Dr.'s Chelsi Brosh and Angel Filer, for providing insights and acting as wonderful listeners to my research excitement. And many thanks to the two wonderful individuals who have served as CEO's of TouchMath during my five years as Chief Academic Officer. Diane Miller, is a great friend and colleague who offered me the opportunity to dive deeply into math and research into how to support students struggling with math. And, to Sean Lockwood, what can you say but many thanks to the person who informs you that "you are going to write a book." Of course, the response was "I'm doing what!" Thank you, Sean, for giving me the time and resources to get this work done and finding out that I love the process of writing.

And many, many thanks to the incredible TouchMath trainers,

especially Dr. Theresa Shattuck and Jenn Hill and the rest of the team of consultants I have been fortunate enough to work with to provide professional development to hundreds of districts and thousands of educators over the last five years. We built a training program that has helped teachers and parents worldwide understand the precepts that make the TouchMath program so effective. You took me down the wonderful path of "why is no one paying attention to dyscalculia" that resulted in this book. And a very special shout out to the first TouchMath trainer I met and have continued to count as a friend, Randy LaRusso who introduced me to the world of dyscalculia, as she has the disability. Her honesty and willingness to share the trials and successes ignited a desire to make a difference in me that will never burn out. Thank you, Randy!

Thank you to my collaborators at TouchMath in the marketing team, including Andrea Black, who kept me on track in the kindest of ways even though the hands in the glove were absolutely made of steel. The gentle but pointed feedback kept me on track and I have learned so much about this process.

I also want to thank Dr. Honora Wall, Dr. Carolyn Stadlman, Dr. Robai Werunga, Amy Smith, Dwight Jones, Laura Drechsel, and all of the educators, parents, and students across the schools, districts, states, and countries that I have had the privilege of working with over the last, dare I say, six decades. I have learned so much from their expertise and their experiences. Thank you.

And of course, my family and friends- parents, siblings, spouse, sons, and grandchildren- who have encouraged me to be what I am and understood my need to write on trains, planes, and whenever. You listened and provided the love and support that mattered.

And lastly, for those whom I did not remember. It is important as we get older to practice senior moments, so that when it is time to have them, we are experts at forgetting names.

References

Abd Halim, F. A., Mohd Ariffin, M., & K. Sugathan, S. (2018). Towards the development of mobile app design model for dyscalculia children in Malaysia. MATEC Web of Conferences, 150, 5016.

Abdou, R. A. E. (2020). The effect of Touch Math multi-sensory program on teaching basic computation skills to young children identified as at risk for the acquisition of computation skills. Amazonia Investiga, 9(27), 149-156.

Afolabi, A. O. (2024). Mathematics Learning through the Lens of Neuroplasticity: A Researcher's Perspective. International Journal of Research and Innovation in Social Science, 8(3s), 4150-4158.

Aksoy ŞG. (2024). The Comorbidity of Specific Learning Disorders in Attention Deficit Hyperactivity Disorder. Comprehensive Medicine, 16(1), 58-62. DOI: 10.14744/cm.2023.30602

Al Otaiba, S., & Petscher, Y. (2020). Identifying and serving udents with learning disabilities, including dyslexia, in the context of multitiered supports and response to intervention. Journal of Learning Disabilities, 53(5), 327-331.

Aldrup, K., Klusmann, U., & Lüdtke, O. (2020). Reciprocal associations between students' mathematics anxiety and achievement: Can teacher sensitivity make a difference? Journal of Educational Psychology, 112(4), 735–750. https://doi.org/j57j

American Psychological Association. (2020). Publication manual of the American Psychological Association (7th ed).

Anobile, G., Bartoli, M., Masi, G., Tacchi, A., & Tinelli, F. (2022). Math difficulties in attention deficit hyperactivity disorder do not originate from the visual number sense. Frontiers in Human Neuroscience, 16, 949391.

Aquil, M. A. I., & Ariffin, M. M. (2020). The causes, prevalence and interventions for dyscalculia in Malaysia. Journal of Educational and Social Research, 10(6), 279.

Aquil, M. A. I. (2020). Diagnosis of Dyscalculia: A Comprehensive Overview. South Asian Journal of Social Sciences and Humanities, 1(1), 43-59.

Altabakhi IW, Liang JW. Gerstmann Syndrome. [Updated 2023 Aug 28]. In: StatPearls [Internet]. Treasure Island (FL): StatPearls Publishing; 2024 Jan-. Available from: https://www.ncbi.nlm.nih.gov/books/NBK519528/

Arizmendi, G. D., Li, J., Van Horn, M. L., Petcu, S. D., & Swanson, H. L. (2021b). Language-Focused interventions on math performance for English learners: A selective Meta-Analysis of the literature. Learning Disabilities Research and Practice, 36(1), 56-75.

Arsalidou, M., & Taylor, M. J. (2011). Is 2+2=4? meta-analyses of brain areas needed forThe numbers and calculations. NeuroImage (Orlando, Fla)., 54(3), 2382-2393.

Ashraf, S., Aftab, M. J., Jahan, M., Bahoo, R., & Altaf, S. (2021). Multicultural education validation of diagnostic test for students with learning difficulties in mathematics at elementary level

Attout, L., & Majerus, S. (2014). Working memory deficits in developmental dyscalculia: The importance of serial order. Informa UK Limited.

Attout, L., & Majerus, S. (2018). Serial order working memory and numerical ordinal processing share common processes and predict arithmetic abilities. Wiley.

Avant, M.J., & Heller, K.W. (2011). Examining the Effectiveness of TouchMath With Students With Physical Disabilities. Remedial and Special Education, 32, 309 - 321.

Bailey, D. H., Fuchs, L. S., Gilbert, J. K., Geary, D. C., & Fuchs, D. (2020). Prevention: Necessary but insufficient? A 2-Year Follow-Up of an effective First-Grade mathematics intervention. Child Development, 91(2), 382-400.

Bailey, D. H., Siegler, R. S., & Geary, D. C. (2014). Early predictors of middle school fraction knowledge. Developmental Science, 17(5), 775-785.

Bandura, A. (2012). On the functional properties of perceived self-efficacy revisited. Journal of Management, 38(1), 9-44.

Barrocas R, Roesch S, Gawrilow C and Moeller K (2020) Putting a Finger on Numerical Development – Reviewing the Contributions of Kindergarten Finger Gnosis and Fine Motor Skills to Numerical Abilities. Front. Psychol. 11:1012. doi: 10.3389/fpsyg.2020.01012

Barroso, C., Ganley, C. M., McGraw, A. L., Geer, E. A., Hart, S. A., & Daucourt, M. C. (2021). A meta-analysis of the relation between math anxiety and math achievement. Psychological Bulletin, 147(2), 134–168.

Baten, E., & Desoete, A. (2018). Mathematical (dis)abilities within the opportunity-propensity model: The choice of math test matters. Frontiers Media SA.

Bergman-Nutley, S., & Klingberg, T. (2014). Effect of working memory training on working memory, arithmetic and following instructions. Psychological Research, 78(6), 869-877.

Berkeley, S. B., Scanlon, D., Bailey, T. R., Sutton, J. C., Sacco, D. M. (2020). A snapshot of RTI implementation a decade later: new picture, same story. Journal of Learning Disabilities, 53(5), 332-342.

Berteletti I and Booth JR (2015) Perceiving fingers in single-digit arithmetic problems. Front. Psychol. 6:226. doi: 10.3389/fpsyg.2015.00226

Biancardi, V. C., & Stern, J. E. (2016). Compromised blood–brain barrier permeability: novel mechanism by which circulating angiotensin II signals to sympathoexcitatory centres during hypertension. The Journal of physiology, 594(6), 1591-1600.

bin Ibrahim, M.Z., Benoy, A. and Sajikumar, S. (2022), Long-term plasticity in the hippocampus: maintaining within and 'tagging' between synapses. FEBS J, 289: 2176-2201.

Bouck, E. C., Satsangi, R., & Park, J. (2018a). The concrete–representational–abstract approach for students with learning disabilities: An evidence-based practice synthesis. Remedial and Special Education, 39(4), 211-228.

Brendefur, J. L., Johnson, E. S., Thiede, K. W., Strother, S., & Severson, H. H. (2018). Developing a multi-dimensional early elementary mathematics screener and diagnostic tool: The primary mathematics assessment. Early Childhood Education Journal, 46(2), 153-157.

Brunner, M., Preckel, F., Götz, T., Lüdtke, O., & Keller, L. (2023). The Relationship Between Math Anxiety and Math Achievement: New Perspectives From Combining Individual Participant Data and Aggregated Data in a Meta-Analysis.

Bryant, D. P., Bryant, B. R., Gersten, R., Scammacca, N., & Chavez, M. M. (2008). Mathematics intervention for first- and second-grade students with mathematics difficulties. SAGE Publications.

Buckley, S. (2020). Mathematics anxiety. Department of Education and Training. https://research.acer.edu.au/learning_processes/28

Bugden, S., Peters, L., Nosworthy, N., Archibald, L., & Ansari, D. (2021). Identifying children with persistent developmental dyscalculia from a 2-min test of symbolic and nonsymbolic numerical magnitude processing. Wiley.

Bugden, S., & Ansari, D. (2014). When your brain cannot do 2 + 2: A case of developmental dyscalculia. Frontiers for Young Minds, 2

Bulat, J., Hayes, A. M., Macon, W., Ticha, R., & Abery, B. H. (2017). School and classroom disabilities inclusion guide for low- and middle-income countries. RTI Press.

Bulthé, J.; Prinsen, J.; Vanderauwera, J.; Duyck, S.; Daniels, Ni.; Gillebert, C.; Mantini, D.; Op de Beeck, H.; De Smedt, Bert. (2018). Multi-method brain imaging reveals impaired representations of number as well as altered connectivity in adults with dyscalculia. NeuroImage. 190.

Burns, M. K., Codding, R. S., Boice, C. H., & Lukito, G. (2010). Meta-analysis of acquisition and fluency math interventions with instructional and frustration level skills: Evidence for a skill-by-treatment interaction. School Psychology Review, 39(1), 69-83.

Butterworth, B. (2018). Dyscalculia: from Science to Education. United Kingdom: Taylor & Francis.

Butterworth, B. (2003). Dyscalculia screener: Highlighting pupils with specific learning difficulties in maths. Nfer Nelson Publishing.

Calik, N.C., & Kargin, T. (2010). Effectiveness of the Touch Math Technique in Teaching Addition Skills to Students with Intellectual Disabilities. International journal of special education, 25, 195-204.

Carey, S. (2009). Where our number concepts come from. The Journal of Philosophy, 106(4), 220-254.

Cheng, D., Xiao, Q., Chen, Q., Cui, J., & Zhou, X. (2018). Dyslexia and dyscalculia are characterized by common visual perception deficits. Developmental neuropsychology, 43(6), 497-507.

Chodura, S., Kuhn, J., & Holling, H. (2015). Interventions for children with mathematical difficulties. Zeitschrift Für Psychologie, 223(2), 129-144.

Cihak, D. F., & Foust, J. L. (2008). Comparing number lines and touch points to teach addition facts to students with autism. Focus on autism and other developmental disabilities, 23(3), 131-137.

Clements, D. H., Sarama, J. (2023). Learning Trajectories in Early Mathematics- Sequences of Acquisition and Teaching. Encyclopedia on Early Childhood Development. Accessed 9.2.24.

Clements, D. H., & Sarama, J. (2021). Learning and teaching early math: The learning trajectories approach. Routledge.

De Smedt, B., Janssen, R., Bouwens, K., Verschaffel, L., Boets, B., and Ghesquière, P. (2009). Working memory and individual differences in mathematics achievement: a longitudinal study from first grade to second grade. J. Exp. Child Psychol. 103, 186–201. doi: 10.1016/j.jecp.2009.01.004

De Visscher, A., Vogel, S. E., Reishofer, G, Hassler, E., Koschutnig, K., De Smedt, B., & Grabner, R. H. (2018). Interference and problem size effect in multiplication fact solving: Individual differences in brain activations and arithmetic performance. NeuroImage, 172, 718-727.

Dehaene, S. (2020). How we learn: why brains learn better than any machine ... for now (First American edition.). Viking.

DeMaria, Kyle (ed.) and Christopher McLaren, "Trends in Disability Employment," Trendlines, U.S. Department of Labor Employment and Training Administration, October 2024, https://www.dol.gov/sites/dolgov/files/ETA/opder/DASP/Trendlines/posts/2024_10/Trendlines_October_2024.html

Dennis, M.S., Sharp, E., Chovanes, J., Thomas, A., Burns, R.M., Custer, B., & Park, J. (2016). A meta-analysis of Empirical research on teaching students with mathematical learning difficulties. Learning Disabilities Research & Practice, 31(3), 156-168. DOI:10.1111/ldrp.12107

Devine, A., Hill, F., Carey, E., & Szűcs, D. (2018). Cognitive and emotional math problems largely dissociate: Prevalence of developmental dyscalculia and mathematics anxiety. Journal of Educational Psychology, 110(3), 431-444.

Dewey, D., Bernier, F.P. The Concept of Atypical Brain Development in Developmental Coordination Disorder (DCD)—a New Look. Curr Dev Disord Rep 3, 161–169 (2016).

DfES (2001). The National Numeracy Strategy. Guidance to Support Learners with Dyslexia and Dyscalculia. London. DfES. 0512/2001

Dinkel, P. J., Willmes, K., Krinzinger, H., Konrad, K., & Koten, J., Jan Willem. (2013). Diagnosing developmental dyscalculia on the basis of reliable single case FMRI methods: Promises and limitations. PloS One, 8(12), e83722.

Doabler, C. T., Clarke, B., Kosty, D., Kurtz-Nelson, E., Fien, H., Smolkowski, K., & Baker, S. (2019). Examining the impact of group size on the treatment intensity of a Tier 2 mathematics intervention within a systematic framework of replication. Journal of Learning Disabilities, 52, 168–180. doi:10.1177/00222194187893

Dowker, A., Sarkar, A., & Looi, C. Y. (2016). Mathematics anxiety: What have we learned in 60 years? Frontiers in Psychology, 7, 508.

Driver, M. K., & Powell, S. R. (2017). Culturally and linguistically responsive schema intervention: Improving word problem solving for English language learners with mathematics difficulty. Learning Disability Quarterly, 40(1), 41-53.

Dueker, S. A., & Day, J. M. (2022). Using standardized assessment to identify and teach prerequisite numeracy skills to learners with disabilities using video modeling. Psychology in the Schools, 59(5), 1001-1014.

Duncan, G. J., Dowsett, C. J., Claessens, A., Magnuson, K., Huston, A. C., Klebanov, P., & Japel, C. (2007). School readiness and later achievement. Developmental psychology, 43(6), 1428.

Durán, R. P., Zhang, T., Sañosa, D., & Stancavage, F. (2020). Effects of Visual Representations and Associated Interactive Features on Student Performance on National Assessment of Educational Progress (NAEP) Pilot Science Scenario-Based Tasks. American Institutes for Research.

Education Endowment Foundation, (2020). Improving math in the early years and key stage 1.

Ellingsen, R., & Clinton, E. (2017). Using the TouchMath Program to Teach Mathematical Computation to At-Risk Students and Students with Disabilities. Educational Research Quarterly, 41, 15-42.

ERIC - Education Resources Information Center. (2022).

Erdem, E., & Gürbüz, R. (2015). An analysis of seventh-grade students' mathematical reasoning. Cukurova University Faculty of Education Journal, 44(1), 123-142.

Espina, E., Marbán, J. M., & Maroto, A. (2022). A retrospective look at the research on dyscalculia from a bibliometric approach. Una mirada retrospectiva a la investigación en discalculia desde una aproximación bibliométrica. Revista de Educación, 396, 201-229.

Faulkenberry, Thomas & Geye, Trina. (2014). The Cognitive Origins of Mathematics Learning Disability: A Review. The Rehabilitation Professional. 22. 9-16.

Fanari, R., Meloni, C., & Massidda, D. (2019). Visual and spatial working memory abilities predict early math skills: A longitudinal study. Front. Psychol. 10:2460. doi: 10.3389/fpsyg.2019.02460

Fengjuan, W., Jamaludin, A. (2023). The Science of Mathematics

Learning: An Integrative Review of Neuroimaging Data in Developmental Dyscalculia. In: Hung, W.L.D., Jamaludin, A., Rahman, A.A. (eds) Applying the Science of Learning to Education. Springer, Singapore. https://doi.org/10.1007/978-981-99-5378-3_3

Fields, R. (2014). Myelin formation and remodeling. Cell, 156(1-2), 15-17.

Fletcher, D., Boon, R. T., & Cihak, D. F. (2010). Effects of the TouchMath program compared to a number line strategy to teach addition facts to middle school students with moderate intellectual disabilities. Education and Training in Autism and Developmental Disabilities, 45(3), 449-458.

Fuchs, L. S., Compton, D. L., Fuchs, D., Paulsen, K., Bryant, J. D., and Hamlett, C. L. (2005). The prevention, identification, and cognitive determinants of math difficulty. J. Educ. Psychol. 97, 493–513. doi: 10.1037/ 0022-0663.97.3.493

Fuchs, L. S., Fuchs, D., Powell, S., Seethaler, P., Cirino, P., & Fletcher, J. (2008). Intensive intervention for students with mathematics disabilities: seven principles of effective practice. (2008). Learning Disability Quarterly, 31(3), 79-92.

Fuchs, L. S., Fuchs, D., & Compton, D. L. (2012). The early prevention of mathematics difficulty: Its power and limitations. Journal of learning disabilities, 45(3), 257-269.

Fyfe, E. R., Matz, L. E., Hunt, K. M., & Alibali, M. W. (2019). Mathematical thinking in children with developmental language disorder: The roles of pattern skills and verbal working memory. Journal of Communication Disorders, 77, 17-30.

Geary, D. C. (2013). Early foundations for mathematics learning and their relations to learning disabilities. Current Directions in Psychological Science : A Journal of the American Psychological Society, 22(1), 23-27.

Geary, D. C. (2011). Cognitive predictors of achievement growth in mathematics: A 5-year longitudinal study. Developmental Psychology, 47(6), 1539-1552.

Gersten, R., Clarke, B. S., Haymond, K., & Jordan, N. C. (2011). Screening for mathematics difficulties in K-3 students. Center on Instruction.

Gilmore, C., Clayton, S., Cragg, L., McKeaveney, C., Simms, V., & Johnson, S. (2018). Understanding arithmetic concepts: The role of domain-specific and domain-general skills. PloS one, 13(9), e0201724.

Gliga, F., & Gliga, T. (2012). Romanian screening instrument for dyscalculia. Procedia - Social and Behavioral Sciences, 33, 15-19.

Grant, J. G., Siegel, Angiulli, A. (2020). From schools to scans: A neuroeducational approach to comorbid math and reading disabilities. Frontiers in Public Health, 8, 469.

Grant, D. (2017). That's the Way I Think: Dyslexia, dyspraxia, ADHD and dyscalculia explained. Taylor & Francis.

Green, N. D. (2009). The Effectiveness of the Touch Math Program with Fourth and Fifth Grade Special Education Students. Online Submission.

Green, C. T., Bunge, S. A., Chiongbian, V. B., Barrow, M., & Ferrer, E. (2017). Fluid reasoning predicts future mathematical performance among children and adolescents. Journal of Experimental Child Psychology, 157, 125-143.

Grigore, M. (2020). Towards a standard diagnostic tool for dyscalculia in school children

Grigorenko, E. L., Compton, D. L., Fuchs, L. S., Wagner, R. K., Willcutt, E. G., & Fletcher, J. M. (2020). Understanding, educating, and supporting children with specific learning disabilities: 50 years of science and practice. American Psychologist, 75(1), 37–51.

Guzmán, B., Rodríguez, C., Ferreira, R. (2022) Effect of parents' mathematics anxiety and home numeracy activities on young children's math performance-anxiety relationship, Contemporary Educational Psychology, Volume 72,102140,ISSN 0361-476X,https://doi.org/10.1016/j.cedpsych.102140

Haberstroh, S., & Schulte-Körne, G. (2019). The diagnosis and treatment of dyscalculia. Deutsches Ärzteblatt International, 116(7), 107-114.

Halberda, J., Mazzocco, M. M., & Feigenson, L. (2008). Individual differences in non-verbal number acuity correlate with maths achievement. Nature, 455(7213), 665-668.

Hattie, J. (2023). Visible learning: The sequel: A synthesis of over 2,100 meta-analyses relating to achievement. Routledge.

Hayes, A. M., Dombrowski, E., Shefcyk, A. H., & Bulat, J. (2018). Learning disabilities screening and evaluation guide for low- and middle-income countries. RTI Press.

Hoedlmoser, K., Peigneux, P., & Rauchs, G. (2022). Recent advances in memory consolidation and information processing during sleep. Journal of sleep research, 31(4), e13607.

Hott, B. L., Morano, S., Peltier, C., Pulos, J., & Peltier, T. (2020). Are students with mathematics learning disabilities receiving FAPE?: Insights from a descriptive review of individualized education programs. Learning Disabilities Research and Practice, 35(4), 170-179.

Hroncich, C. (2022). NAEP 2022: Gloomy Results on the Nation's Report Card.

Hughes, C. A., Morris, J. R., Therrien, W. J., & Benson, S. K. (2017). Explicit Instruction: Historical and Contemporary Contexts. Learning Disabilities Research & Practice, 32(3), 140-148.

Huijsmans, M. D. E., Kleemans, T., van der Ven, S. H. G., & Kroesbergen, E. H. (2020). The relevance of subtyping children with mathematical learning disabilities. Research in Developmental Disabilities, 104, 103704. https://10.1016/j.ridd.2020.1037

Institute of Education Sciences (IES) Home Page, a part of the U.S. Department of Education. (2022). Ed.gov.

Izard V, Dehaene-Lambertz G, Dehaene S (2008) Distinct cerebral pathways for object identity and number in human infants. PLoS Biol 6(2): e11. doi:10.1371/journal. pbio.0060011

Jacobsen, K. H. (2020). Introduction to health research methods: A practical guide. Jones & Bartlett Publishers.

Jaya, A. A. (2009). Development and standardization of dyscalculia screening tool

Jordan, N., Hanich, L. B., & Uberti, H. Z. (2013). Mathematical thinking and learning difficulties. In The development of arithmetic concepts and skills (pp. 383-406). Routledge.

Kankaraš, M., Montt, G., Paccagnella, M., Quintini, G., & Thorn, W. (2016). Skills matter: Further results from the survey of adult skills. OECD skills studies. OECD Publishing. Retrieved from ERIC

Kaufmann, L., Mazzocco, M. M., Dowker, A., von Aster, M., Göbel, S. M., Grabner, R. H., & Nuerk, H. C. (2013). Dyscalculia from a developmental and differential perspective. Frontiers in psychology, 4, 516.

Kim, S. A., Bryant, D. P., Bryant, B. R., Shin, M., & Ok, M. W. (2022). A Multilevel Meta-Analysis of Whole Number Computation Interventions for Students With Learning Disabilities. Remedial and Special Education, 07419325221117293.

Kisler, C., Schwenk, C., & Kuhn, J. (2021). Two dyscalculia subtypes with similar, low comorbidity profiles: A mixture model analysis. Frontiers in Psychology, 12, 589506.

Kohn, J., Rauscher, L., Kucian, K., Käser, T., Wyschkon, A., Esser, G., & von Aster, M. (2020). Efficacy of a computer-based learning program in children with developmental dyscalculia. what influences individual responsiveness? Frontiers in Psychology, 11, 1115.

Kong, J. E., Arizmendi, G. D., & Doabler, C. T. (2022). Implementing the science of math in a culturally sustainable framework for students with and at risk for math learning disabilities. Teaching Exceptional Children, XX(X), 4005992211273.

Kot, M., Terzioğlu, N., Aktaş, B., & Yikmiş, A. (2018). Effectiveness of Touch Math technique: Meta-anaysis study. European Journal of Special Education Research (3)4. doi:http://dx.doi.org/10.46827/ejse.v0i0.1847

Kroesbergen, E. H., Huijsmans, M. D. E., & Kleemans, M. A. J. (2022). The heterogeneity of mathematical learning disabilities: Consequences for research and practice. International Electronic Journal of Elementary Education, 14(3), 227-241.

Kucian, K., & von Aster, M. (2015). Developmental dyscalculia. European journal of pediatrics, 174, 1-13.

Kucian, K., Grond, U., Rotzer, S., Henzi, B., Schönmann, C., Plangger, F., & von ASTER, M. (2011). Mental number line training in children with developmental dyscalculia. NeuroImage, 57(3), 782-795.

Kuhl, U., Sobotta, S., & Skeide, M. A. (2021). Mathematical learning deficits originate in early childhood from atypical development of a frontoparietal brain network. PLoS Biology, 19(9), e3001407.

Kurth, J. A., & Jackson, L. (2022). Introduction to the Special Issue on the Impact of Placement on Outcomes for Students with Complex Support Needs. Research and Practice for Persons With Severe Disabilities, 47(4), 187-190.

Lewis, K. E., Thompson, G. M., & Tov, S. A. (2021). Screening for characteristics of dyscalculia: Identifying unconventional fraction understandings. International Electronic Journal of Elementary Education, 14(3), 243-267.

Litkowski, E. C., Duncan, R. J., Logan, J. A. R., & Purpura, D. J. (2020). Alignment between children's numeracy performance, the kindergarten common core state standards for mathematics, and state-level early learning standards. AERA Open, 6(4), 233285842096854.

Lopes-Silva, J. B., Moura, R., Júlio-Costa, A., Wood, G., Salles, J. F., & Haase, V. G. (2016). What is specific and what is shared between numbers and words? Frontiers in Psychology, 7, 22.

Lynn, A., Wilkey, E. D., & Price, G. R. (2021). Predicting children's math skills from task-based and resting-state functional brain connectivity. Cerebral Cortex (New York, N.Y. 1991), 32(19), 4204-4214.

Mahmud, M. S., Zainal, M. S., Rosli, R., & Maat, S. M. (2020). Dyscalculia: What we must know about students' learning disability in mathematics? Universal Journal of Educational Research, 8(12B), 8214-8222.

Martin, B. N., & Fuchs, L. S. (2022). Predicting risk for comorbid reading and mathematics disability using Fluency-Based screening assessments. Learning Disabilities Research and Practice, 37(2), 100-112.

Mattison, R. E., Woods, A. D., Morgan, P. L., Farkas, G., & Hillemeier, M. M. (2023). Longitudinal trajectories of reading and mathematics achievement for students with learning disabilities. Journal of Learning Disabilities, 56(2), 132-144.

Mazzocco, M. M., Feigenson, L., & Halberda, J. (2011). Impaired acuity of the approximate number system underlies mathematical learning disability (dyscalculia). Child Development, 82(4), 1224-1237.

McCaskey, U., von Aster, M., O'Gorman Tuura, R., & Kucian, K. (2017). Adolescents with developmental dyscalculia do not have a generalized magnitude deficit – processing of discrete and continuous magnitudes. Frontiers in Human Neuroscience, 11, 102.

McCoy, D. C., Salhi, C., Yoshikawa, H., Black, M., Britto, P., & Fink, G. (2018). Home-and center-based learning opportunities for preschoolers in low-and middle-income countries. Children and Youth Services Review, 88, 44-56.

Melby-Lervåg, M., Redick, T. S., & Hulme, C. (2016). Working memory training does not improve performance on measures of intelligence or other measures of "far transfer" evidence from a meta-analytic review. Perspectives on Psychological Science, 11(4), 512-534.

Menon, V., Padmanabhan, A., & Schwartz, F. (2021). Cognitive neuroscience of dyscalculia and math learning disabilities.

Mononen, R., Aunio, P., Koponen, T., & Aro, M. (2014). A review of early numeracy interventions for children at risk in mathematics. International Journal of Early Childhood Special Education, 6(1), 25.

Morsanyi, K., van Bers, B. M., McCormack, T., & McGourty, J. (2018). The prevalence of specific learning disorder in mathematics and comorbidity with other developmental disorders in primary school-age children. British Journal of Psychology, 109(4), 917-940.

Mueller, M., Yankelewitz, D., & Maher, C. (2012). A framework for analyzing the collaborative construction of arguments and its interplay with agency. Educational Studies in Mathematics, 80, 369-387.

Mulchay, Chris & Wolff, Michael & Ward, Julian & Han, Nicole. (2022). Test Review of the Feifer Assessment of Mathematics (FAM). Journal of Pediatric Neuropsychology. 8. 10.1007/s40817-022-00128-y.

National Center for Education Statistics. (2024). Students With Disabilities. Condition of Education. U.S. Department of Education, Institute of Education Sciences. Retrieved [date], from .

National Center for Education Statistics. (2022). Students with disabilities. NCES.ed.gov.

National Institutes of Health (NIH). (2022). National Institutes of Health (NIH). National Institutes of Health (NIH).

NAEP Nations Report Card - National Assessment of Educational Progress - NAEP. (2022). Ed.gov; National Center for Education Statistics.

NAEP Nations Report Card-National Assessment of Educational Progress-NAEP. (2024). Ed. Gove; National Center for Education Statistics. https://www.nationsreportcard.gov/highlights/mathematics/2022/

Nelson, J.A. (2019). An Investigation of the Effectiveness of TouchMath on Mathematics Achievement for Students with the Most Significant Cognitive Disabilities, ProQuest LLC, Ed.D. Dissertation, Kansas State University.

Nelson, G., Crawford, A., Hunt, J., Park, S., Leckie, E., Duarte, A., Brafford, T., Ramos-Duke, M., & Zarate, K. (2022). A systematic review of research syntheses on students with mathematics learning disabilities and difficulties. Learning Disabilities Research and Practice, 37(1), 18-36.

New York State Education Department. (2017). NYSED.gov.

Nicolson, R. I., & Fawcett, A. J. (2021). Mathematics disability vs. learning disability: A 360 degree analysis. Frontiers in Psychology, 12, 725694.

Nosworthy, N., Bugden, S., Archibald, L., Evans, B., & Ansari, D. (2013). A two-minute paper-and-pencil test of symbolic and nonsymbolic numerical magnitude processing explains variability in primary school children's arithmetic competence. PloS one, 8(7), e67918.

Nuffield Foundation. (2018). Dyscalculia, a hidden condition, could be the real reason some people struggle with math. Accessed 5.21.24

Padmanabhan, A., & Schwartz, F. (2017). The oxford handbook of developmental cognitive neuroscience (1st ed).. Oxford University Press.

Park, J., Ramirez, G., & Park, D. (2024). Effect of preschool teacher's math anxiety on teaching efficacy and classroom engagement in math. Psychology in the Schools, 61, 2600–2611.

Peters, L., de Beeck, H. O., & De Smedt, B. (2020). Cognitive correlates of dyslexia, dyscalculia and comorbid dyslexia/dyscalculia: Effects of numerical magnitude processing and phonological processing. Research in Developmental Disabilities, 107, 103806.

Powell, S. R., Fuchs, L. S., & Fuchs, D. Reaching the Mountaintop: Addressing the Common Core Standards in Mathematics for Students with Mathematics Difficulties. Learning Disabilities Research & Practice, 28(1), 38-48.

Prado, J. (2020). The effect of math anxiety on the numerical brain.

Prescott, J., Gavrilescu, M., & Egan, G.F. (2010). Enhanced brain connectivity in math-gifted adolescents: An fMRI study using mental rotations . Cognitive Neuroscience, 4(1)

Price, G., & Ansari, D. (2013). Dyscalculia: Characteristics, causes, and treatments. Numeracy, 6(1), 2.

Purpura, D. J., & Logan, J. A. (2015). The nonlinear relations of the approximate number system and mathematical language to early mathematics development. Developmental Psychology, 51(12), 1717.

Räsänen, P., Aunio, P., Laine, A., Hakkarainen, A., Väisänen, E., Finell, J., Rajala, T., Laakso, M., & Korhonen, J. (2021). Effects of gender on basic numerical and arithmetic skills: Pilot data from third to ninth grade for a large-scale online dyscalculia screener. Frontiers in Education (Lausanne), 6

Rays, B. J. (1994). Promoting number sense in the middle grades. Mathematics Teaching in the Middle School, 1(2), 114-120.

Riley-Tillman, T. C., Burns, M. K., & Kilgus, S. P. (2020). Evaluating educational interventions: Single-case design for measuring response to intervention. Guilford Publications.

Santos, F. H., Ribeiro, F. S., Dias-Piovezana, A. L., Primi, C., Dowker, A., & von Aster, M. (2022). Discerning developmental dyscalculia and neurodevelopmental models of numerical cognition in a disadvantaged educational context. Brain Sciences, 12(5), 653.

Sasanguie, D., De Smedt, B., & Reynvoet, B. (2017). Evidence for distinct magnitude systems for symbolic and non-symbolic number. Psychological research, 81, 231-242.

Schaeffer, M.W., Rozek, C.S., Maloney, E.A., Berkowitz, T., Levine, S.C. and Beilock, S.L. (2021), Elementary school teachers' math anxiety and students' math learning: A large-scale replication. Dev. Sci., 24: e13080.

Szczygieł, M., & Pieronkiewicz, B. (2021). Exploring the nature of math anxiety in young children: Intensity, prevalence, reasons. Mathematical Thinking and Learning, 24(3), 248–266. https://doi.org/10.1080/10986065.2021.1882363

Siemann, J., & Petermann, F. (2018). Evaluation of the triple code model of numerical processing—Reviewing past neuroimaging and clinical findings. Research in Developmental Disabilities, 72, 106-117.

Simon, R., & Hanrahan, J. (2004). An evaluation of the Touch Math method for teaching addition to students with learning disabilities in mathematics. European journal of special needs education, 19(2), 191-209.

Soares, N., Evans, T., & Patel, D. R. (2018). Specific learning disability in mathematics: A comprehensive review. Translational Pediatrics, 7(1), 48-62.

Swanson, H.L., Lussier, C., & Orosco, M. (2013). Effects of cognitive strategy interventions and cognitive moderators on word problem solving in children at risk for problem solving difficulties. Learning Disabilities Research & Practice, 28(4), 170-183.

Szardenings, C., Kuhn, J., Ranger, J., & Holling, H. (2018). A diffusion model analysis of magnitude comparison in children with and without dyscalculia: Care of response and ability are related to both mathematical achievement and stimuli. Frontiers in Psychology, 8, 1615.

Taneja, K. K. and Sankhian, A. (2019). Effect of multi-sensory approach on performance in mathematics at primary level. The Educational Beacon, 8, 93-101.

Träff, U., Olsson, L., Östergren, R., & Skagerlund, K. (2016). Heterogeneity of developmental dyscalculia: Cases with different deficit profiles. Frontiers in Psychology, 7, 2000.

Urton, K., Grünke, M., & Boon, R. T. (2022). Using a Touch Point Instructional Package to Teach Subtraction Skills to German Elementary

Students At-Risk for LD. International Electronic Journal of Elementary Education, 14(3), 405-416.

U.S. Department of Education. (2017). Every Student Succeeds Act (ESSA). U.S. Department of Education.

U.S. Department of Education. (2020). Every Student Succeeds Act (ESSA). U.S. Department of Education.

Üstün, S., Ayyıldız, N., Kale, E. H., Mançe Çalışır, Ö, Uran, P., Öner, Ö, Olkun, S., & Çiçek, M. (2021). Children with dyscalculia show hippocampal hyperactivity during symbolic number perception. Frontiers in Human Neuroscience, 15, 687476.

Vandecruys, F., Vandermosten, M., & De Smedt, B. (2024). The role of formal schooling in the development of children's reading and arithmetic white matter networks. Developmental Science, e13557.

van Bergen, E., de Zeeuw, E., Hart, S. A., Boomsma, D., de Geus, E., & Kan, K. J. (2023, July 26). Comorbidity and causality among ADHD, dyslexia, and dyscalculia. https://doi.org/10.31234/osf.io/epzgy

Van Luit, J. E. H., & Toll, S. W. M. (2018). Associative cognitive factors of math problems in students diagnosed with developmental dyscalculia. Frontiers in Psychology, 9, 1907.

Van Viersen, S., Slot, E. M., Kroesbergen, E. H., Van't Noordende, J. E., & Leseman, P. P. M. (2013). The added value of eye-tracking in diagnosing dyscalculia: A case study. Frontiers in Psychology, 4, 679.

Viesel-Nordmeyer, N., Ritterfeld, U., & Bos, W. (2021). Acquisition of mathematical and linguistic skills in children with learning difficulties. Frontiers Media SA.

Vigna, G., Ghidoni, E., Burgio, F., Danesin, L., Angelini, D., Benavides-Varela, S., & Semenza, C. (2022). Dyscalculia in early adulthood: Implications for numerical activities of daily living. Brain Sciences, 12(3), 373.

Vinson, B. M. (2004). A foundational research base for the touchmath program. Retrieved from Touchmath Website: http://www. TouchMath. com.

Vinson, B. M. (2005). Touching points on a numeral as a means of early calculation: does this method inhibit progression to abstraction and fact recall. Athens State University.

Von Aster M. (2000). Developmental cognitive neuropsychology of number processing and calculation: varieties of developmental dyscalculia. European child & adolescent psychiatry, 9 Suppl 2, 1141–1157.

Von Aster, M. G., & Shalev, R. S. (2007). Number development and developmental dyscalculia. Developmental medicine & child neurology, 49(11), 868-873.

Wagner, R. K., & Torgesen, J. K. (1987). The nature of phonological processing and its causal role in the acquisition of reading skills. Psychological Bulletin, 101(2), 192.

Wall, H. (2022). Teaching Students with Dyscalculia. PolyMath Publishing.

Walsh, M. M., Krusmark, M. A., Jastrembski, T., Hansen, D. A., Honn, K. A., & Gunzelmann, G. (2023). Enhancing learning and retention through the distribution of practice repetitions across multiple sessions. Memory & Cognition, 51(2), 455-472.

Waters, H. E., & Boon, R. T. (2011). Teaching Money Computation Skills to High School Students with Mild Intellectual Disabilities via the TouchMath© Program: A Multi-Sensory Approach. Education and Training in Autism and Developmental Disabilities, 46(4), 544–555.

Watson, S. M., Gable, R. A., & Morin, L. L. (2016). The role of executive functions in classroom instruction of students with learning disabilities. International Journal of School and Cognitive Psychology, 3(167).

Wells, A. (1997). Does numeracy matter? Adults Learning, 8(6), 151-152.

Willcutt EG, Petrill SA, Wu S, Boada R, Defries JC, Olson RK, Pennington BF. Comorbidity between reading disability and math disability: concurrent psychopathology, functional impairment, and neuropsychological functioning. J Learn Disabil. 2013 Nov-Dec;46(6):500-16. doi: 10.1177/0022219413477476.

Wilkey, E. D., & Ansari, D. (2020). Challenging the neurobiological link between number sense and symbolic numerical abilities. Annals of the New York Academy of Sciences, 1464(1), 76-98.

Williams, K., White, S.L.J. & English, L.D. Profiles of general, test, and mathematics anxiety in 9- and 12-year-olds: relations to gender and mathematics achievement. Math Ed Res J (2024).

Wilson, A. J., & Dehaene, S. (2007). Number sense and developmental dyscalculia. In D. Coch, G. Dawson, & K. W Fischer (Eds.), Human behavior, learning, and the developing brain: Atypical development (pp. 212–238). The Guilford Press.

Wilson, A. (2016). Dyscalculia Primer and Resource Guide - OECD. www.oecd.org. https://www.oecd.org/education/ceri/dyscalculiaprimerandresourceguide.htm

Wisniewski, Z. G., & Smith, D. (2002). How Effective Is Touch Math for Improving Students with Special Needs Academic Achievement on Math Addition Mad Minute Timed Tests?.

Yikmis, A. (2016). Effectiveness of the touch math technique in teaching basic addition to children with Autism. Educational Sciences:Theory & Practice, 16, 1005-1025.

Yoong, S. M., Hosshan, H., Arumugam, S., Alya Qasdina Ng Ai Lee, Lau, S. C., & Govindasamy, P. (2022). Validity and reliability of needs analysis questionnaire for dyscalculia instrument. South Asian Journal of Social Science and Humanities, 3(3), 111-124.

Zacharopoulos, G., Sella, F., & Kadosh, R. C. (2021). The impact of a lack of mathematical education on brain development and future attainment. PNAS Proceedings of the National Academy of Sciences of the United States of America, 118(24), Article e2013155118.

Zakariya, Y.F. (2022). Improving students' mathematics self-efficacy: A systematic review of intervention studies. Frontiers in Psychology, 13.

Zhang, Sunming. (2022). Math Anxiety: The Influence of Teaching Strategies and Teachers' Attitude.

Made in the USA
Monee, IL
16 March 2025